Numbers and infinity

Numbers and infinity

A historical account of mathematical concepts

ERNST SONDHEIMER

AND

ALAN ROGERSON

Westfield College, University of London

CAMBRIDGE UNIVERSITY PRESS

Cambridge

London   New York   New Rochelle

Melbourne   Sydney

Published by the Press Syndicate of the University of Cambridge
The Pitt Building, Trumpington Street, Cambridge CB2 1RP
32 East 57th Street, New York, NY 10022, USA
296 Beaconsfield Parade, Middle Park, Melbourne 3206, Australia

First published 1981

Printed in Great Britain at the University Press, Cambridge

Library of Congress catalogue card number: 81-7660

British Library Cataloguing in Publication Data

Sondheimer, Ernst
Numbers and infinity.
1. Mathematics – 1961-
I. Title II. Rogerson, Alan
510   QA36
ISBN 0 521 24091 3 hard covers
ISBN 0 521 28433 3 paperback

CONTENTS

PREFACE

In this book we treat mathematical ideas in a historical setting. Our principal themes are the concepts of *number* (Chapters 1–7) and *infinity* (Chapters 8–11). Within this context we have selected a number of topics for discussion with an eye to their history. Mathematical knowledge of roughly A-level standard is presupposed, and the book is addressed primarily to mathematics students in the sixth form at school and the first year at university who have an interest in historical aspects of their subject. The reader is presented with mathematical argument from time to time so that he may rise above the level of generalities and anecdotes. We hope that this will not deter any intelligent layman who wishes to find out about the development of ideas in mathematics. The more technical passages can simply be skipped or taken for granted.

The book differs both from conventional histories of mathematics and from mathematics textbooks. The history of mathematics is a vast, fascinating and difficult subject, much of it of specialised interest and of little direct concern to the mathematician of today. Many excellent accounts exist (see the bibliography for some recommended titles). We are not here primarily interested in an account of 'what happened in history', with its many blind alleys and side-issues, but we have tried to give some indication of the way men's ideas about mathematics have changed through time. In a good modern mathematics course the subject is, quite properly, developed logically on the basis of the most up-to-date views of its nature. There is usually little time to question or discuss the underlying assumptions and methods, and the complex evolution of mathematical ideas over the ages tends to be little mentioned. The student is thus presented with a beautiful and impressive logical structure.

This may give him the impression that the assumptions made, and the methods used, in handling basic concepts like number, function, continuity, should be quite obvious to him. (Only a fool would think otherwise!) In fact many of the ideas about the nature of mathematics which are taken for granted today are of very recent origin, and it may reassure the student, as he grapples with concepts that he finds difficult and that are not at all 'obvious' to him, to know that these concepts were also not obvious to people in the past, and that it took much argument, confusion and searching for the 'right' answer before the modern approach was created. Thus some knowledge of the historical background to modern mathematics should help to elucidate modern mathematics itself.

Rarely indeed were new concepts discovered all at once or immediately accepted. In the case of complex numbers (involving $\sqrt{(-1)}$) there was no particular year in which someone 'discovered' or 'invented' the idea. These numbers (which now seem so basic and clear cut) hovered on the fringe of mathematics for hundreds of years, half-understood, argued about, accepted (more or less) by some, rejected by others. Nor is the development of ideas in a discipline necessarily a smooth continuous evolutionary process. It is subject to long periods of non-development, to sudden jumps or changes in understanding and even at times to 'backward' steps. If the ideas of Popper, Kuhn and Lakatos are of interest to the reader, he will find in their works stimulating views of the progress of scientific thought, all of which see development as being essentially discontinuous. It must of course be remembered that the contemporary writer is looking back with hindsight and that his judgements are contemporary ones. He selects the results that are now considered to be significant. A historical account of mathematical ideas written in the twentieth century will inevitably reflect what twentieth-century mathematicians think is important in mathematics. But this must not lead the historical writer, when he describes the work of authors of the past, into ascribing to *them* the modern knowledge and the modern ways of thinking on which his own judgements are based. He has thus a difficult task: he must at one and the same time try to understand the way people thought in the past and evaluate the evidence in the light of his knowledge of the present.

Our style is deliberately informal, and we proceed by example rather than proof. Mathematicians will not find our book 'respectable' in matters of definition; existence and uniqueness are usually taken for granted. Thus in Chapter 7, where we talk about hypercomplex numbers, we unashamedly treat notions such as 'space', 'rotation', 'dimension', as intuitively given. In modern mathematics these words receive a precise technical definition and become more general and abstract than the everyday notions from which they arose. This is a necessary process from which mathematics derives its power and flexibility, and the point has not been ignored in our text. But there is often a loss of contact between the mathematician's abstract definitions and the scientist's need to use similar concepts in situations of practical interest; sometimes the problem is just one of confusion of language when the specialists in different disciplines try to communicate. We strongly believe that our education system should aim to produce both mathematically educated scientists and mathematicians able to apply their knowledge. If our book does a little towards bridging existing gaps we shall be delighted.

Having said all this we must also admit that there is much arbitrariness in our choice of topics. The field of mathematics is far too vast for everything of interest to be mentioned in a work of modest compass which attempts a treatment that is not purely superficial. For the same reason we have, usually, resisted the temptation to take the story into the twentieth century – or sometimes even far into the nineteenth. These limitations were also enforced by our restriction to elementary mathematics, and (not least) by our own mathematical ignorance. We sincerely hope, however, that no reader will think that we are only talking about old issues which are dead-and-buried today. Many of the topics treated are still very much alive and can furthermore be tackled much more powerfully with modern mathematical technique. We have tried to indicate this in a few places, but the elementary nature of our book has prevented us from emphasising the point as much as it deserves. The mathematically knowledgeable reader will be only too aware of all that we have omitted. (A parallel book could, and should, be written on geometry, for instance.) Finally, at our level of treatment we claim no original insights. Nor have we made direct use of primary sources, although we have tried to avoid perpetuating popular misconceptions about the history of mathematics. Our indebtedness to other authors will be obvious from a reading of the books recommended in the bibliography. We have taken what we needed from these excellent secondary sources without explicit acknowledgement on each occasion. Our thanks and apologies are tendered herewith.

The book grew out of a lecture course given (by E.S.) to first year undergraduates at Westfield College. The students were expected to participate by writing essays. This brings us to another comment on mathematics teaching. Mathematics students when they go into careers in management and administration often find that they are handicapped because their education has not trained them in the skill of English writing. A study of the history of mathematics provides an opportunity to remedy this. A list of suggested essay topics is included in the appendix, but the reader will have no difficulty in thinking of other titles.

A number of colleagues have helped with this book, either with specific suggestions or with general encouragement. In particular, our thanks go to Bryan Thwaites for encouraging us to turn the lecture notes into a book, to Patricia Bennett, Geoffrey Howson and Eira Scourfield for a critical reading of the entire manuscript and to Janet Sondheimer for help with the preface. The authors of course accept full responsibility for all errors and defects remaining in the text.

1

Representation of numbers

Let us look at the way we write numbers. Take whole numbers (positive integers) first, for example the number of days in a normal year, a number which we write as 365. This is quite a subtle way of writing such a large number. It uses a *decimal system* with *base* 10, and *place values*: thus 5 measures the 'units', 6 the 'tens' and 3 the 'hundreds', so that 365 stands for

$$3 \times 10^2 + 6 \times 10^1 + 5 \times 10^0.$$

The arrangement of the digits matters: 563 is quite a different number from 365. We need 9 digits $(1, 2, \ldots, 9)$ together with the zero (the symbol for an empty place), and these are sufficient to write any number. The zero symbol is needed, for example, to distinguish 35 from 305. This so-called 'Hindu–Arab' system of writing numbers became established in the West during the late Middle Ages, gradually displacing an earlier more primitive Roman system which can still be seen on some old clocks.

Let us consider some other ways of writing numbers. The most primitive system uses notches in a tally stick or the equivalent: knots on a string, heaps of pebbles, and so on. Here 12 appears as |||||||||||| , or perhaps in bundles: ||||| ||||| || . In this system it is a laborious task to carve out the notches which represent the number of days in a year. Such methods were used by primitive tribes and may still be used nowadays by a scorekeeper in a ball game. The earliest known tally stick is the radius bone of a wolf which has 55 notches, the first 25 in groups of 5. It dates from the Old Stone Age, more than 10 000 years ago. Bundles were usually heaps of 5, 10 or 20, presumably because counting was done with the aid of the fingers on one hand, on both hands, or with fingers and toes; later special symbols were introduced to represent 5, 10, etc. Our modern ciphers 2 and 3 are believed to have originated from the primitive $=$ and $\equiv$ by connecting the horizontal lines in cursive writing.

Primitive number systems of this kind were used for thousands of years. The most familiar to us is the Roman system, an additive decimal system with special signs for the higher units, but without the notion of place values. Thus 1867 is written as MDCCCLXVII, which stands for

$$1000 + 500 + (3 \times 100) + 50 + 10 + 5 + (2 \times 1).$$

Note that V stands for half of X, D for half of $\bigoplus$ which is the old symbol for 1000. The system was very awkward for computation: try multiplying the above number by LXVII using only Roman numerals. It needed experts to do such sums.

Long before the time of the Romans, mathematics developed in the civilisations of the Ancient Orient: Egypt, Mesopotamia (Babylon), China and India. The study and elucidation of the mathematics of these ancient times is the business of archaeologists. A great deal has been discovered quite recently, although certain knowledge is hard to obtain; there is much argument about the correct interpretation of the sources (see Neugebauer 1969 for a detailed account). The Egyptian *hieroglyphic* numeration is found in carvings on tombs and monuments, and some of it is 5000 years old (as old as the pyramids). It uses a decimal system similar to the Roman, with separate symbols for powers of ten (thus 10 is ∩, a heel bone, 100 is ೨ , a scroll, 1000 is ⚘, a lotus flower, and so on). Using this system the Egyptians found it easy to write numbers as large as a million. For Egyptian mathematics of a later period the sources are papyrus records such as the famous *Rhind papyrus* in the British Museum. This papyrus was discovered in 1858 and is known to have been written about 1650 B.C. (but it contains older material). The Rhind papyrus is not written in hieroglyphics but in a running script called *hieratic* (or 'sacred', to distinguish it from a still later *demotic* or 'popular' script). The hieratic numerals make an important advance in that they introduce the use of special signs to represent the digits from 1 to 9: thus 8 is no longer represented by repetition of the unit: $\frac{||||}{||||}$, but by the single symbol $=$.

From Babylonia we have well-preserved clay tablets with mathematical texts. These have revealed a level of calculating skill much higher than the Egyptian. As early as 2000 B.C. the Babylonians used a number system based on the number 60 (a 'sexagesimal' system) which, remarkably enough, incorporated the place value idea: thus 4,2,7 would stand for our number

$$4 \times 60^2 + 2 \times 60^1 + 7 \times 60^0 = 14\,527.$$

This ancient Babylonian system was in essentials equivalent to our own. In some ways it was even better: there are advantages in using 60 as a base rather than 10, because 60 has many more integer factors and thus subdivisions are more easily made. Of course the Babylonians did not use our modern symbols to write their numbers: their number symbols were 'cuneiform' (wedge-shaped), with $\curlyvee$ standing for 1, $\curlyvee\curlyvee$ for 2, $\langle$ for 10 and $\langle\!\!\langle\!\!\langle\substack{\curlyvee\curlyvee\curlyvee\\\curlyvee\curlyvee\curlyvee\\\curlyvee\curlyvee\curlyvee}$ for 59. The Babylonian system was well adapted to computation, especially because the place-value idea was extended to include (sexagesimal) fractions as well as integers; and the Babylonians were able to achieve remarkable accuracy in numerical approximation. Thus the square root of 2 was obtained (in modern symbols) as

$$1, 24, 51, 10 = 1 + \frac{24}{60} + \frac{51}{60^2} + \frac{10}{60^3} = 1.414\,212\,9\ldots$$

(correct value $1.414\,213\,5\ldots$). Such accuracy was not surpassed until the time of the Renaissance, over 3000 years later.† Powerful as the system was, the interpretation of Babylonian

† How did the Babylonians calculate square roots? They used the following powerful method – now known as 'Newton's algorithm' – to obtain their approximations: to find $\sqrt{a}$, suppose a_1 is a first approximation; form $b_1 = a/a_1$ and $a_2 = \frac{1}{2}(a_1 + b_1)$ (the second approximation); then form $b_2 = a/a_2$, $a_3 = \frac{1}{2}(a_2 + b_2)$ (the third approximation); and so on. (This is actually an infinite process, but was not recognised as such in ancient times.) If $a = \alpha^2 + \beta$ and $a_1 = \alpha$, this method gives $b_1 = \alpha + (\beta/\alpha)$ and $a_2 = \alpha + (\beta/2\alpha)$, the two-term approximation to the binomial series for $\sqrt{(\alpha^2 + \beta)}$.

numbers is not unambiguous, because there was no 'decimal point' to separate whole numbers from fractions, and (in the earlier periods) there was not yet a specific symbol denoting zero (so that an empty place *might* denote a zero): thus 4,2,7 could mean

$$4 \times 60^1 + 2 \times 60^0 + 7 \times 60^{-1}, \text{ or } 4 \times 60^2 + 2 \times 60^0 + 7 \times 60^{-1}.$$

The correct interpretation was presumably to be determined by the context in which the number appeared. The later Babylonians had a zero symbol for an empty space in the interior of a number, but not for a zero at the end.

In spite of much discussion the real reason for the Babylonians's introduction of a sexagesimal system is not definitely known. It appears to have arisen in response to the needs of their highly developed administrative and commercial activities which called for a convenient and accurate way of doing numerical calculations. Both the place-value principle and the sexagesimal system have remained with us permanently; the number base 60 appears in our own division of the hour into 60 minutes and 60^2 seconds, and of the circle into 6×60 degrees, each degree into 60 minutes and each minute into 60 seconds. In fact measurements of time intervals and angles were important for the later Babylonians (after 700 B.C.) who developed a scientific system of astronomy in which they kept systematic records of the positions of the sun, moon and planets.

The Greeks developed abstract pure mathematics to a high level but did not do much calculating with numbers. So perhaps we should not be surprised that they had a more primitive number system which had no place values. The early form of the Greek system was rather like the Roman and the Egyptian. Later, in Alexandrian times, a better system appears in which computation was not too difficult. This used 27 Greek letters (nine for $1, 2, \ldots, 9$; nine for $10, 20, \ldots, 90$; nine for $100, 200, \ldots, 900$); thus one could write any number less than 1000 with at most three symbols. Such a system was used by scientists, merchants and administrators for about 15 centuries.

Most of the Greek mathematical texts were in fact not written in Greece but in the Egyptian city of Alexandria during the period 350–200 B.C., well after the 'Golden Age' of Greece, around 430 B.C. Alexandria had become the centre of 'Hellenistic' Greek civilisation after the conquests of Alexander the Great. Euclid, the best-known Greek mathematician, probably lived in Alexandria at about 300 B.C. but nothing certain is known about his life. In fact we have virtually no direct sources to give us a picture of the development of Greek mathematics. We do have reliable editions of the work of the great Hellenistic mathematicians, Euclid, Archimedes and Apollonius. These give a very detailed account of Greek mathematics, but are texts systematically covering a highly developed mathematical science in which historical origins are hard to trace – much like most modern textbooks. So we have to reconstruct the formative period of Greek mathematics from various fragments transmitted by later authors, and from scattered remarks made by philosophers and others. Thus Plato, although not himself a mathematician, was very interested in mathematics and often talked about it and the achievements of mathematicians. A vast amount of research (see Heath 1921) has allowed us to construct a fairly consistent, but largely hypothetical, picture of the way Greek mathematics developed.

Under the Roman Empire Alexandrian mathematics combined the Greek tradition of pure abstract mathematics with the computational arithmetic and algebra of the Egyptians and Babylonians. For example, Ptolemy's famous treatise on astronomy, the *Almagest* (Arabic for 'Great Collection'), written about A.D. 150, contained a trigonometry, with tables of chords for different angles (equivalent to our sine tables). Chords were expressed as sexagesimal fractions, thus

$$\sin 1° \text{ was } 1,2,50 = \frac{1}{60} + \frac{2}{60^2} + \frac{50}{60^3} = 0.017\,453\,7\ldots$$

(actual value $0.017\,452\,4\ldots$).

With the decline of the Roman Empire the focus of mathematical activity shifted eastwards to India, then back to Mesopotamia (now under the Arabs). It is the achievement of Indian mathematics to have combined the ancient notions of a decimal system and of a place-value (position) system into our modern method of writing numbers. Hindu astronomers, around A.D. 500, used words to represent digits ('moon' to denote 1, 'fire' or 'brothers' to denote 3, etc.); these were arranged in a positional system starting with units, then tens, hundreds, etc. Thus our number 365 would actually appear (in word form) as 563. Tables of sines could thus be prepared and memorised in verse form! Some time soon after A.D. 500 the Hindus changed to a digital notation which included the symbol o for zero (adopted from the Greeks), and switched to the Greek–Babylonian order for the numerals in which the largest units were written first. Hindu arithmetic then looked much like our own, but the Hindu numerals spread to Europe only very gradually, by way of the Arabs who translated the Indian texts.† Arab mathematics flourished in the centuries following the year 800. One of the best-known authors was Al-Khowarizmi (about 825) who wrote many books on mathematics and astronomy. A Latin translation of his arithmetic appeared about 1100 and served to popularise the Indian number system in Western Europe. Many modern mathematical terms are of Arab origin: thus 'algorithm' (meaning a systematic method for computation) is a latinisation of Al-Khowarizmi's name. His *Al-jabr wa'l-muqabalah* (science of reduction and cancellation) gave us the word 'algebra'; originally it meant the science of equations. The word 'zero' comes from the Latin 'zephirum' which is derived from the Arabic 'sifr' (sifra = empty); hence also the English word 'cipher'.

Over the period A.D. 1000–1500 the Hindu–Arab number system co-existed in Western Europe with the Greek and Roman numerals. Oriental science was brought to the West by Italian merchants who travelled in the East. One of them was Leonardo of Pisa (Fibonacci, 'son of Bonaccio') who wrote a famous mathematical treatise (in 1202), the *Liber Abaci*, which was influential in introducing Arabic numerals to the West. However, on the whole, people preferred the Roman numerals with which they were familiar; they had learned to do sums with them quite rapidly using an abacus (a counting board with

---

† The detailed circumstances of the introduction of Hindu–Arab numerals into Europe are still disputed among historians of mathematics.

movable counters, still in use in some parts of the world). The public disliked Hindu–Arab numerals because they were strange and difficult to read, and the authorities opposed them because they were too easily forged. In 1299 Florentine merchants were forbidden to use Arabic numerals in book-keeping; 200 years later Roman numerals had disappeared entirely from the books of the Medici.

Why should we use the Hindu–Arab system to represent numbers? Computation (addition and multiplication) is easy in this system; also we can tell at a glance whether one number is greater than another. (This is not true in the Roman system: compare C and LXX for example.) These advantages come from the place-value idea, but they do not depend on using the number 10 as a base. There is no particular mathematical reason for the use of this base. Earlier on, as we have noted, very accurate calculations had been made with the sexagesimal system (for example by Ptolemy). We can use any other base and still have the same advantages. In the seventeenth century the mathematician Weigel (who taught Leibniz; see Chapter 9) spent much of his life advocating the use of a number system with base 4 (nobody took much notice). Leibniz himself pointed out (in 1703) that in the *binary system*, with base 2, the digits 0 and 1 were sufficient to represent any number, but he did not recommend it for practical use as the expressions were too long. Nowadays the binary system is fundamental for modern computers.

How do we write the number 365 in binary form? We must express it as a sum of powers of 2. We have $2^1 = 2$, $2^2 = 4$, $2^3 = 8$, $2^4 = 16$, $2^5 = 32$, $2^6 = 64$, $2^7 = 128$, $2^8 = 256$, ... The largest power of 2 less than 365 is 2^8, and $365 - 2^8 = 109$; the largest power of 2 less than 109 is 2^6; and so on. Thus

$$\begin{aligned}
365 &= 2^8 + 109 = 2^8 + 2^6 + 45 \\
&= 2^8 + 2^6 + 2^5 + 13 \\
&= 1 \times 2^8 + 0 \times 2^7 + 1 \times 2^6 + 1 \times 2^5 + 0 \times 2^4 \\
&\quad + 1 \times 2^3 + 1 \times 2^2 + 0 \times 2^1 + 1 \times 2^0,
\end{aligned}$$

and the binary form of 365 is therefore 101 101 101. In practice,

to obtain the binary form of any number, we repeatedly divide
by 2 and note the remainders:

$$
\begin{array}{rl}
2)\underline{365} & \\
2)\underline{182} & \text{r. 1} \\
2)\underline{\ 91} & \text{r. 0} \\
2)\underline{\ 45} & \text{r. 1} \\
2)\underline{\ 22} & \text{r. 1} \\
2)\underline{\ 11} & \text{r. 0} \\
2)\underline{\ 5} & \text{r. 1} \\
2)\underline{\ 2} & \text{r. 1} \\
2)\underline{\ 1} & \text{r. 0} \\
0 & \text{r. 1}
\end{array}
$$

The remainders in *ascending* order give 101 101 101, the correct
binary expression for 365. This systematic way of obtaining the
binary expression for any number always works (can you see
why?). Try it with a few other numbers.

It is true that the binary form of a number is much longer than
the decimal, but there are important advantages:

(i) we need only two digits 0, 1 to represent any number:
hence base 2 is suitable for electronic computers which work
with on/off circuit elements;

(ii) we have very simple addition and multiplication tables,
as follows:

+	0	1		×	0	1
0	0	1		0	0	0
1	1	10		1	0	1

Thus, to multiply 365 by 13, we convert 13 to its binary form
1101, and proceed by long multiplication:

$$
\begin{array}{r}
101101101 \\
\times \quad\ 1101 \\
\hline
101101101 \\
101101101\ \ \\
101101101\ \ \ \\
\hline
1001010001001
\end{array}
$$

(Notice the binary addition sums used here: $1+1+1=11$, $1+1+1+1=10+1+1=11+1=100$.) Check that the answer is the same as the decimal result: $365 \times 13 = 4745$. This arithmetic seems strange at first because it is unfamiliar, but it is in principle very easy. Multiplication involves only addition and 'moving to the left': a simple job for computers.

We have only considered whole numbers so far; let us now have a look at *fractions*. What do we mean when we represent a fraction such as $\frac{3}{8}$ by its decimal equivalent, 0.375? The decimal form 0.375 is an extension of our previous ideas on place values using base 10: 0.375 means

$$3 \times 10^{-1} + 7 \times 10^{-2} + 5 \times 10^{-3}.$$

This is another way of writing the fraction $\frac{3}{8}$. Notice that in order to obtain this decimal representation we need only divide 3 by 8:

```
       0.375
    8)3.0000
      2.4
       .60
       .56
       .040
       .040
       . . .
```

What happens, however, when we try to find the decimal form of the simple fraction $\frac{2}{3}$? Try it and see. You will find that this simple fraction produces an *infinite* (or 'periodic') decimal (one that never stops):

$$\frac{2}{3} = 0.6666\ldots = 6 \times 10^{-1} + 6 \times 10^{-2} + 6 \times 10^{-3} + \cdots$$

This is in fact an infinite series:

$$\frac{6}{10} + \frac{6}{100} + \frac{6}{1000} + \cdots$$

Does this make sense? If this infinite series is to represent the fraction $\frac{2}{3}$ then by adding up *all* its terms we should get $\frac{2}{3}$; thus $\frac{2}{3}$ should be equal to the *sum* of the infinite series. How do we find such a sum? We obtain it as the *limit*, as N tends to infinity (we write $N \to \infty$), of the sum s_N of the first N terms of the series. This limit indeed has the value $\frac{2}{3}$, i.e., as N gets larger and larger, s_N gets closer and closer to $\frac{2}{3}$. To check this statement we find the sum of N terms of our series (this is easy because we have a *geometric series*) and then let $N \to \infty$:

$$s_N = \frac{6}{10}\left(1 - \frac{1}{10^N}\right) \Big/ \left(1 - \frac{1}{10}\right)$$

$$= \frac{2}{3}\left(1 - \frac{1}{10^N}\right) \to \frac{2}{3} \text{ as } N \to \infty$$

(since $1/10^N$ tends to zero).

Are things better in another base? Try base 2. (A ruler divided into halves, quarters, eighths, ... of an inch uses base 2 for fractions.) Check that the fraction $\frac{2}{3}$ is still represented by an infinite decimal (0.101 01 ...). There is an interesting confirmatory proof to show that this binary expansion cannot be finite. If it were finite, we would have

$$\frac{2}{3} = \sum_{n=1}^{m} \frac{a_n}{2^n},$$

a finite sum terminating with $a_m/2^m$, say. Hence we would have $\frac{2}{3} = A/2^m$, with A some integer, or $2^{m+1} = 3A$: but we know that 2^{m+1} is not divisible by 3.

We get a simpler 'decimal' if we use base 3: $\frac{2}{3} = 0.2$; but now we find that $\frac{1}{2}$ in base 3 is an infinite 'decimal': 0.1111.... Thus no particular base is best for all fractions. We will however show in Chapter 3 that the 'decimal' form of all fractions, in base 10 or any other base, is either *finite* or *periodic*. A periodic (infinite) decimal has a group of digits which repeats. (See Rademacher & Toeplitz 1957, ch. 23, for a discussion of the length of the period and related matters.)

The Egyptians, about 4000 years ago, had a remarkable calculus of *unit fractions*. The idea was to write all fractions as sums of fractions with unit numerator and distinct denominators, for example

$$\frac{3}{23} = \frac{1}{10} + \frac{1}{46} + \frac{1}{115}.$$

This is quite ingenious but is very awkward for calculation. Also note that such a representation is not unique; thus

$$\frac{3}{5} = \frac{1}{2} + \frac{1}{10} = \frac{1}{3} + \frac{1}{5} + \frac{1}{15} = \frac{1}{3} + \frac{1}{4} + \frac{1}{60}.$$

It is thought that because of this cumbersome calculus Egyptian mathematics failed to progress beyond a primitive stage.

How can we construct this representation systematically? The following technique was described by Fibonacci in the *Liber Abaci*. Take the fraction $\frac{3}{13}$ as an example. Find the *largest* unit fraction $\leqslant \frac{3}{13}$: this can be done by increasing the denominator of the given fraction until it first reaches a multiple of the numerator. We obtain $\frac{3}{15} = \frac{1}{5}$. Subtract this from $\frac{3}{13}$:

$$\frac{3}{13} - \frac{1}{5} = \frac{15 - 13}{65} = \frac{2}{65}.$$

Now repeat the process: the largest unit fraction $\leqslant \frac{2}{65}$ is $\frac{2}{66} = \frac{1}{33}$. Subtract:

$$\frac{2}{65} - \frac{1}{33} = \frac{66 - 65}{2145} = \frac{1}{2145}.$$

At this stage we are left with a unit fraction, and thus

$$\frac{3}{13} = \frac{1}{5} + \frac{1}{33} + \frac{1}{2145}.$$

It is quite easy to show that this always works, that the process stops after a finite number of steps. Suppose A/B is a fraction between 0 and 1, with $A < B$, and $1/N$ is the largest unit fraction $\leqslant A/B$, then we are to form

$$\frac{A}{B} - \frac{1}{N} = \frac{NA - B}{BN},$$

and (if necessary) to repeat the process. (Of course, if A divides B, there is nothing to prove.) We need only show that the numerator of the new fraction, $NA - B$, is *less* than the original numerator: $NA - B < A$ (try to show this!) Then the numerators, which by construction are positive, diminish at each step, and must thus decrease to zero in a finite number of steps. Thus every positive fraction between 0 and 1 can be represented as a finite sum of unit fractions.

This process, however, is not the one used by the ancient Egyptians. The Rhind papyrus has tables listing decompositions into unit fractions, but $\frac{2}{15}$ for example is split into $\frac{1}{10} + \frac{1}{30}$, whereas our process gives $\frac{1}{8} + \frac{1}{120}$. There has been much speculation about the procedures used by the Egyptians to obtain their particular representations, but we do not really know what they were.

Egyptian unit fractions suggest more interesting (and harder) mathematical problems. Suppose for example that we limit ourselves to fractions with *odd* denominators only. Can such fractions always be written as sums of distinct unit fractions with odd denominators? (For example $\frac{2}{9} = \frac{1}{5} + \frac{1}{45}$.) The answer is yes, but the proof given above now fails because the numerators no longer necessarily decrease, and it is not known in this case whether Fibonacci's process of generating the unit fractions always stops after a finite number of steps.

The decimal fractions we use nowadays are much more suitable for numerical calculations. They were generally adopted in Europe around 1600: a book published by Simon Stevin, a Flemish engineer, in 1585 (*De Thiende*, meaning 'The Tenth') was influential in making them widely known. Interest in methods of computation was widespread at that time: engineers, astronomers, surveyors were all looking for better ways to do arithmetical calculations. Symptomatic of the search for greater accuracy was the effort to calculate π to as many decimal places as possible: in 1610 Ludolph van Ceulen computed 35 decimal places by Archimedes's method of inscribed and circumscribed polygons (see p. 97). He spent much of his life on the task and it was regarded as a great triumph of computation.

The introduction of logarithms soon after 1600 gave a powerful impetus to computational technique. In 1614 John Napier, the Laird of Murchiston in Scotland, published his famous work *Mirifici Logarithmorum Canonis Descriptio.*† ·(Jobst Bürgi in Switzerland had similar ideas at about the same time.) Napier's basic idea was to reduce multiplication to addition through the use of two series of numbers related in such a way that, when one grows in arithmetic progression, the other decreases in geometric progression. There is then a simple relation between the product of two numbers in the second

† 'A Description of the Marvellous Rule of Logarithms'. In the seventeenth century Latin was still, as it had been in the Middle Ages, the international language for scientific writing.

series and the sum of the corresponding numbers in the first series. This general idea was not new, but to put it into practice it was necessary to define a system of logarithms that was both convenient and accurate for numerical work. Napier's logarithm was defined as follows: $x = \text{Naplog } y$ if (in modern notation) $y = 10^7 e^{-x/10^7}$; thus $\text{Naplog } y = 10^7 (\log 10^7 - \log y)$, where $\log y$ is our 'natural logarithm' to base e. In this system, if $\text{Naplog } y = \text{Naplog } y_1 + \text{Naplog } y_2$, y is actually not $y_1 y_2$ but $y_1 y_2 / 10^7$.[†] Napier himself was not satisfied with his rather clumsy definition, and it was left to his admirer Henry Briggs, professor at Gresham College in London, to publish in 1624 (after Napier's death) the first tables of 'Briggsian logarithms' which are our common logarithms based on $y = 10^x$. The new technique was immediately recognised as a most important aid for simplifying complicated calculations, and tables of logarithms were soon produced which satisfied all practical requirements.

The discovery of logarithms stimulated the development of mechanical aids to calculation. The ancient abacus, useful enough for day-to-day commercial transactions, was too crude a tool for doing complicated scientific calculations. Napier himself invented a device of rectangular strips ('Napier's rods' or 'Napier's bones') which helped in carrying out long multi-plications. This was popular in its time, but much more important was the invention of the slide rule by the Reverend William Oughtred in 1622. This was a device in which two pieces of wood marked with logarithm scales can slide relative to each other; by adding logarithmic distances one is multiply-ing the associated numbers. Up to the arrival, a few years ago, of pocket electronic calculators a slide rule was the standard calculating aid of every scientist and engineer. The inventor of the first widely known *calculating machine* was Blaise Pascal (see p. 117) who in 1642 designed and manufactured an instru-ment with interlocking dials which could perform the 'carrying'

[†] Napier's definitions were equivalent to these but he was not familiar with the concepts of the exponential function or of a basis for logarithms. Clarification of these ideas had to wait for the introduction of the calculus (see Chapter 9).

process in addition and could thus add numbers mechanically. Leibniz invented an improved machine which could also multiply and divide automatically. Later, in the early nineteenth century, the 'irascible genius' Charles Babbage spent his life designing much more elaborate calculating machines. His ambitious projects were never completed, but Babbage's 'analytical engine' contained many of the features (such as punched cards, a memory, input and output devices) of the modern electronic computer. The development of today's computers dates from the Second World War. As processors of information of all kinds they are revolutionising modern technology; as tremendously powerful calculating devices they are having a profound influence on mathematics. Goldstine (1972) gives an interesting account of the history of the computer.

A final remark: we have in this chapter considered only integers and fractions, i.e. *rational numbers*. Here, as we have noted, the decimal representation is either finite or, if infinite, is periodic. Numbers such as $\sqrt{2}$ or π, which we shall consider later, have no decimal representation of this simple type. They can be written as infinite decimals, but no pattern can be seen to generate all the digits. The non-mathematician may say π is 3.1416 (or 22/7, etc.). These rational numbers are certainly good numerical *approximations* for π, and this is usually sufficient for the practical scientist or engineer concerned with measuring. But for the mathematician π is quite a different sort of number which we shall consider in more detail in later chapters.

2

The integers

In this chapter we shall discuss a few of the properties of the *natural numbers* (positive integers) 1, 2, 3, ... As objects of mathematical study they have been of great interest since Greek times. In earlier days, number lore had been associated with magic and astrology: 'mathematici' in ancient terminology meant magicians or astrologers. The word 'mathematics' itself comes from a Greek root, and means 'something learned, science'. An example of number magic is 'gematria', a form of Hebrew cabbalistic mysticism in which each letter of the alphabet has a number value. Thus texts could be used to prophesy: in Isaiah (21:8) the lion proclaims the fall of Babylon because the letters in the Hebrew words for lion and Babylon add up to the same sum. While astrology has survived, mathematicians are no longer interested in such arguments.

The mathematical study of the integers is called the 'theory of numbers'. It is a branch of mathematics which has few practical uses. It is also fascinating, because many of the problems are both easy to formulate and very hard to answer. Gauss (1777–1855) – one of the greatest mathematicians – said 'Mathematics is the queen of the sciences, and the Theory of Numbers is the queen of mathematics'.

Let us consider the question of *divisibility*. A number which is composite can be written as a product of at least two other numbers different from 1, for example $6 = 2 \times 3$. (2, 3 are the *divisors* of 6.) *Prime numbers* have no divisors except 1 and the number itself. The first few primes are

$$2, 3, 5, 7, 11, 13, 17, 19, 23, 29, 31, 37, \ldots,$$

not counting 1 as a prime number. Clearly 2 is the only *even* prime. Prime numbers are the basic integers because every

15

composite integer can be written as a product of primes (thus $120 = 2 \times 2 \times 2 \times 3 \times 5 = 2^3 \times 3 \times 5$). The prime numbers are therefore the building bricks for the construction of all the integers. As mathematicians we are interested in general questions about the class of prime numbers, and not so much in the properties of any particular prime. Look at the list of prime numbers. There is no obvious regularity, but a few questions suggest themselves immediately. For example:

1. How many primes are there?
2. How big can the gaps be between successive primes?
3. How many pairs of primes differ by 2 ('twin primes', such as 5, 7; 17, 19)?
4. Can we find a *formula* $f(n)$ to represent the nth prime?

Some questions of this type are fairly easy to answer, others can be answered but are difficult, and yet others are so hard that we do not know the answer (and may not even know where to look for the answer). Many other interesting questions about the integers could be listed, and the theory of numbers, which has been studied since ancient times, is today a highly active field of research for many of the world's leading mathematicians. (See Dudley 1978 for a good elementary account of the subject.) Let us look at the above questions one by one.

1. *How many primes are there?* This was answered by the Greeks: *there is no end to the primes*. The proof given in Book 9 of Euclid is a classic example of a mathematical proof. Look at the following numbers:

$$2 \times 3 + 1 = 7,$$
$$2 \times 3 \times 5 + 1 = 31,$$
$$2 \times 3 \times 5 \times 7 + 1 = 211,$$
$$2 \times 3 \times 5 \times 7 \times 11 + 1 = 2311,$$
$$2 \times 3 \times 5 \times 7 \times 11 \times 13 + 1 = 30031, \text{ etc.}$$

(To form them we multiply together the first n primes, $n = 2, 3, 4, \ldots$, and add 1.) We see that none of these numbers is divisible by any of the primes used to form it, because there is always a remainder 1. This does *not* however mean that the numbers themselves are necessarily prime: what it *does* mean is that each number is *either* prime, *or*, if it is not, then its prime factors must be greater than the prime numbers used in the construction. In fact 7, 31, 211, 2311 are all primes, but $30\,031 = 59 \times 509$. Thus this process *generates new primes* from a collection of the first n primes. This shows that there is no 'last prime'. Why? Because, if p were to be the last prime, we form $2 \times 3 \times 5 \times 7 \times \cdots \times p + 1 = N$; then our argument shows that N is either prime or has prime factors bigger than p. Thus p cannot be the last prime.

This is an elegant proof, simple and tailored to the problem in hand. It does not attempt to do the 'obvious' thing, which would be to look for the *next* prime after p. That question presents a much harder problem because there is no simple law for the sequence of successive primes. Euclid's proof instead looks for *some* prime beyond p, and this is all that is needed for the purpose of the proof. Asking the question in the right way is often the secret of success in mathematics.

2. *How big can the gaps be between successive primes*? This question was not considered by the Greeks. We can show that the gaps between successive primes can be as large as we please; there is no upper limit. Given any number N, we can always find N consecutive composite numbers. For example to show that there are 99 consecutive composite numbers, we need only form the number $100! = 1 \times 2 \times 3 \times 4 \times \cdots \times 100$, and consider the 99 numbers

$$100! + 2, \quad 100! + 3, \quad 100! + 4, \ldots, \quad 100! + 100.$$

These are clearly all composite, since, for any number k from 2 to 100, $100!$ is divisible by k and therefore $100! + k$ is also divisible by k (note that $100! + 1$ might be prime). Of course these are very large numbers, so we have to go far along the sequence of primes to get large gaps, but evidently we can get gaps as large as we please by going far enough.

An interesting related question is: do arithmetical series of integers like 1, 4, 7, 10, ... or 3, 7, 11, 15, ... contain an infinity of primes? The answer is yes, but Euclid's type of proof works only for special sequences (try to construct a proof for 3, 7, 11, 15, ...). The general proof given by Dirichlet in 1837 is difficult and uses analytic (i.e. calculus) methods. (See Davenport 1970 for further details.)

3. *How many pairs of primes differ by 2?* This question looks innocent enough but is extremely hard to answer. It is believed that there are infinitely many twin primes but nobody knows how to prove it. The conjecture arose from looking at the sequence of primes and noting that twins such as 5, 7; 11, 13; 17, 19 keep occurring however long the list. Modern computers have immensely increased our ability to investigate properties of primes by this kind of 'experimental' approach.

A related and famous unsolved problem is the *Goldbach conjecture* (1742). Goldbach, who was the Prussian envoy to Russia, stated in a letter to the leading mathematician Euler that *every even number is the sum of two primes* (thus $8 = 5 + 3$, $16 = 13 + 3$, $48 = 29 + 19$, ...). The assertion that this is true for *every* even number was made without proof, and it is Goldbach's only claim to mathematical fame. (Nowadays interesting conjectures are still often made, but only a high-powered mathematician is likely to put forward a new conjecture which is worthy of serious attention.) No exceptions to the Goldbach conjecture are known, but it has never been proved. Properties such as this one which relate primes to *addition* are hard to prove since the primes are defined by multiplication. There was no progress with the Goldbach problem at all until 1931, when a young Russian mathematician showed that every positive integer can be represented as the sum of not more than N primes, where N was a large number around 300 000. Since then there has been more progress and it is now known that every *sufficiently large* even number can be written in the form $p + q$, where p is prime and q has at most 2 prime factors. It is often easier to demonstrate a result as a *limiting* property for numbers that are large enough than to give a proof which holds precisely for every number. The next question illustrates this strikingly.

4. *Can we find a formula $f(n)$ which, when we substitute in it the integer n, gives us the nth prime number?* People searched for such a formula in the seventeenth and eighteenth centuries but were unsuccessful. No simple formula exists; the distribu-

tion of primes is too irregular. Suppose we are less ambitious and look for a function $f(n)$ which produces *only* primes, but need not give them all. Fermat (1601–65), the chief originator of the modern theory of numbers, conjectured that all numbers of the form $f(n) = 2^{2^n} + 1$ $(n = 1, 2, 3, \ldots)$ are prime. Check that

$$f(1) = 5, \quad f(2) = 17, \quad f(3) = 257, \quad f(4) = 65\,537;$$

all these numbers are indeed prime numbers. Clearly these 'Fermat numbers' increase very rapidly. It is not at all easy to determine whether any given very large number is prime or not, and it was a considerable feat of computation for his time when Euler showed in 1732 that $f(5)$ is composite: in fact

$$f(5) = 2^{32} + 1 = 4\,294\,967\,297 = 641 \times 6\,700\,417.$$

Thus, while a conjecture may seem plausible on the basis of a few examples, we must not believe it until a satisfactory mathematical *proof* has been given. With modern computers we can investigate numbers which are vastly larger than Euler's $f(5)$: at the time of writing (summer 1980) the largest number which is definitely *known* to be a prime number is the 'Mersenne number' $2^{44\,497} - 1$, a large number indeed!

But there are limits to what even the best present-day computers can do. Suppose we form a number of say 400 arabic digits by multiplying together two large primes, each of about 200 digits; then no computer, presented with such a product, could in practice disentangle the factorisation; the calculations required would simply take far too long. This fact underlies the recent invention of a system of coding messages called 'public-key cryptography'. The basic idea is to use the 400-digit number to encode the message, and to have a rule for decoding which can be operated only by someone who knows the prime factors. Since anyone who does not know the factors has no way of finding them within his lifetime, the code effectively unbreakable. Thus prime numbers have their uses after all! But mathematicians are hard at work trying to discover more powerful 'factoring algorithms', so the security of the code may be short-lived. (For fuller details see Hellman 1979, Kolata 1980.)

Even if a formula for the nth prime can be written down, it will not necessarily be of any use for computation in practice. In recent years formulae have actually been given from which the $(n+1)$th prime can be obtained when the first n primes are known; but they are very complicated. In the nineteenth century progress was made by formulating the problem in a different way. It was noticed that the law governing the *distribution* of primes appears to have a fairly simple form, at least for sufficiently large numbers. Thus, instead of searching for a formula for the nth prime, one looks for the *average* distribution of the primes among the integers.

Suppose $F(n)$ is the number of primes among the first n integers: thus $F(10) = 4$ (the primes $\leqslant 10$ are 2, 3, 5, 7). As n tends to infinity $F(n)$ increases also without limit, since there are infinitely many primes, but it increases more slowly than n. The *density* of primes among the first n integers is $F(n)/n$. This is also the *probability* that an integer picked at random from the first n integers is a prime number. When this number is computed it is found that it *decreases* as n increases, which means that the average *gap* between primes increases slowly (as we might expect):

n	$F(n)$	$F(n)/n$	$1/\log n$	$\dfrac{F(n)/n}{1/\log n}$
10^3	168	0.168	0.145	1.159
10^6	78 498	0.078	0.072	1.084
10^9	50 847 478	0.051	0.048	1.053

We now look for a function $f(n)$ which 'behaves similarly', at least for sufficiently large n. We want a function which decreases slowly with increasing n in much the same way as $F(n)/n$. Legendre and Gauss noticed (about 1800) that the simple function $1/\log_e n$ has the required behaviour. This is seen from the table above, where the ratio of $F(n)/n$ to $1/\log n$ appears to approach 1 as $n \to \infty$. The conjecture was thus made that the ratio indeed tends to the limit 1 as $n \to \infty$, or:

$$F(n) \sim \frac{n}{\log n} \left(\text{this means} \lim_{n \to \infty} \frac{F(n)}{n/\log n} = 1 \right).$$

It is a surprising relation because there is no obvious connection between the logarithmic function and the prime numbers. It took about 100 years before the first complete proofs of the conjecture were given (by Hadamard and de la Vallée Poussin in 1896); these proofs established the relation as the *prime number theorem*. The proofs used analytical methods, involving calculus and complex function theory. This was not the end of the story. Various refinements of the theorem were made giving better approximations, and it was also felt by many mathematicians that one should be able to derive results concerning only the integers by using proofs which operate *only* with integers and which do not use any 'foreign ideas' such as analysis. An 'elementary' proof of the prime number theorem was given by Selberg and Erdös in 1948; it succeeded in using only the integers, but it was by no means a *simple* proof. Very recently a much simpler proof has been given (see Newman 1980) which does however use some basic complex analysis. What constitutes a 'good' proof in mathematics is a matter open to argument and to changes in fashion.

Next we say something about the question of *unique factorisation* of a number into prime factors. It is easy to show that any integer can be written as a product of primes, but can this be done in only one way? Thus we can factorise the number 60 as $6 \times 10 = (3 \times 2) \times (2 \times 5) = 2^2 \times 3 \times 5$, or alternatively as $4 \times 15 = 2^2 \times 3 \times 5$, which gives the same factors. It may seem obvious that we should always get the same result, but suppose we take the Fermat number

$$2^{32} + 1 = 4\,294\,967\,297 = 641 \times 6\,700\,417;$$

is it really obvious that this is the *only* factorisation into primes? The unique factorisation property of the integers is a result that has to be proved; it is called the *fundamental theorem of arithmetic*. Euclid does not state it explicitly (hence many school books tend to assume the proof is unnecessary), but he does give a result which is essentially equivalent: if p (a prime) divides the product ab, then it must divide either a or b. The fundamental theorem is easily derived from this.

It becomes evident that unique factorisation is a property which needs proof when we construct other number systems for which the property does not hold. Consider, for example, the system which consists only of integers of the form $3n + 1$, i.e. the numbers $1, 4, 7, 10, 13, \ldots$. It is easy to check that the product of any two such numbers is also of the form $3n + 1$ and thus belongs to the system. Further, whilst the number 16 factorises as 4×4, the numbers 4, 10, 25 have no factors within the system and are thus primes in the system. Moreover, we have $100 = 4 \times 25 = 10 \times 10$, and thus we have two *different* representations of the number 100 as products of prime factors!

Another example where uniqueness fails is obtained when we consider all numbers of the form $a + b\sqrt{6}$, where a and b are positive or negative integers. Such generalisations of the integers are called *algebraic numbers* because they are roots of algebraic equations with integer coefficients: thus $2 + \sqrt{6}$ is a root of $x^2 - 4x - 2 = 0$. Algebraic numbers may be *complex*: $2 + \sqrt{(-6)}$ is a root of $x^2 - 4x + 10 = 0$. (Complex numbers are discussed in Chapter 3: here all we need to know is that $\sqrt{(-6)} \times \sqrt{(-6)} = -6$.) The numbers $a + b\sqrt{6}$ include the integers as the special case $b = 0$. Note that addition and multiplication of these numbers always gives numbers of the same class: thus

$$(3 + \sqrt{6}) + (-2 + \sqrt{6}) = 1 + 2\sqrt{6},$$
$$(3 + \sqrt{6})(-2 + \sqrt{6}) = -6 - 2\sqrt{6} + 3\sqrt{6} + 6$$
$$= \sqrt{6}, \text{ etc.}$$

It is not immediately clear which numbers are primes in this system. If we consider the factorisation of the number 6, there are apparently two different forms, since we can write $6 = 2 \times 3$, and also $6 = \sqrt{6} \times \sqrt{6}$. But the factors 2, 3, $\sqrt{6}$ are actually *not* primes in our system since we have

$$\sqrt{6} = (3 + \sqrt{6})(-2 + \sqrt{6}) = (2 + \sqrt{6})(3 - \sqrt{6}),$$

and

$$2 = (2 + \sqrt{6})(-2 + \sqrt{6}), \quad 3 = (3 + \sqrt{6})(3 - \sqrt{6}).$$

So in fact both the above factorisations of the number 6 lead to the same result:

$$6 = (2 + \sqrt{6})(-2 + \sqrt{6})(3 + \sqrt{6})(3 - \sqrt{6}),$$

and these four factors *are* the prime factors of the number 6 in our system. Thus it turns out that here unique factorisation still applies, although the result was hardly obvious. And, if we now consider the system of complex algebraic numbers $a + b\sqrt{(-6)}$, then we have the two factorisations $6 = 2 \times 3 = (-\sqrt{(-6)}) \times (\sqrt{(-6)})$, and now it turns out that 2, 3, $\sqrt{(-6)}$ *are* primes in the system and cannot be factorised. So here we again have two different factorisations into prime factors.

The theory of algebraic numbers is a large subject, developed intensively in the nineteenth century. A leading originator was Kummer (1810–93) who participated in the discussions on the validity (or otherwise) of unique factorisation. The question was of interest in the 1840s in connection with efforts to find a general proof of *Fermat's last theorem*. This arose as follows: the equation $a^2 + b^2 = c^2$ can be solved for a class of positive integers; the simplest example is $a = 3$, $b = 4$, $c = 5$, and in general all numbers of the form $a = u^2 - v^2$, $b = 2uv$, $c = u^2 + v^2$, with u, v positive integers and $u > v$, satisfy the equation. Such numbers are called *Pythagorean triples* because they represent (in accordance with the theorem of Pythagoras) the sides of a right-angled triangle in which the lengths of the sides are *commensurable* (i.e. the ratios of the lengths of the sides are ratios of whole numbers, see p. 33). The general method for constructing Pythagorean triples was given by the Greek mathematician Diophantos who lived about A.D. 250 in Alexandria; his *Arithmetica* contains the first systematic use of algebraic symbols (but a modern reader will not find it easy to follow his notation). In the margin of his copy of Diophantos's work Fermat wrote (in Latin): 'However, it is impossible to write a cube as the sum of two cubes, a 4th power as the sum of two 4th powers, and in general any power beyond the 2nd as the sum of two similar powers. For this I have discovered a truly wonderful proof, but the margin is too small to contain it.'

Thus Fermat's last theorem asserts that $a^n + b^n = c^n$ has no solution in positive integers a, b, c if n is an integer greater than 2. It is really a conjecture rather than a theorem because, despite all efforts over more than 300 years, it has never been proved as a general result valid for all n, though its truth has been demonstrated for every single value of n from 3 up to a few thousand. Some special cases are fairly easy to deal with; the proof for

$n = 4$ was given by Fermat himself. Fermat's theorem became a famous problem which, while not in itself very important, led to new developments in number theory and also attracted many amateurs. A prize of 100 000 marks was at one time offered in Germany for a proof, but its value was wiped out by the 1923 inflation. It is the arithmetical problem for which the greatest number of incorrect 'proofs' have been given!

Complex algebraic numbers arise when we factorise $a^n + b^n$ into linear factors:

$$a^n + b^n = (a+b)(a+rb)(a+r^2b) \cdots (a+r^{n-1}b),$$

where r is a complex number such that $r^n = 1$. (For example, when $n = 3$, $r = -\frac{1}{2} + \frac{1}{2}\sqrt{(-3)}$.) Attempts to prove Fermat's theorem in the 1840s failed because they assumed unique factorisation into primes for such complex numbers. This raises the question: can we define special kinds of algebraic numbers for which unique factorisation again holds? This *can* be done (they are called 'ideals'), and pursuit of this idea by Kummer and later mathematicians led to a powerful generalisation of the simple notion of arithmetic divisibility with many applications in number theory and modern algebra. The theory of ideals allowed Kummer to prove Fermat's last theorem for a large class of prime number exponents n called 'regular primes' (but not for all values of n).

Fermat's problem is an example of a *Diophantine equation*, an equation to be solved with integral values for the unknowns. Such equations lead to some very difficult problems in the theory of numbers. However, the simplest case, *the linear Diophantine equation in two unknowns*,

$$ax + by = c, \tag{1}$$

where a, b, c are given natural numbers and x, y are the unknowns, can be dealt with quite easily. Note that such an equation might have no solutions at all, or a finite number of solutions, or an infinite number. We shall derive the necessary and sufficient condition for equation (1) to have a solution in (positive or negative) integers.

For this purpose we introduce the number (a, b), the *greatest common divisor* of a and b, i.e. the largest integer which is a factor of both a and b. Then (a, b) has the interesting property that integers k, l always exist such that

$$(a, b) = ka + lb; \qquad (2)$$

we say that (a, b) is *linearly dependent* on a and b. This result follows from a systematic process, given by Euclid, for finding (a, b), the *Euclidean algorithm*. We illustrate it by means of an example. Let $a = 432$, $b = 156$; we are to express 432 as a multiple of 156 with a remainder:

$$432 = 2 \times 156 + 120.$$

(The idea is that any integer which divides 432 and 156 also divides 120; hence the common divisors of 432 and 156 are the same as the common divisors of the smaller pair 156 and 120.) Now we continue the process: we express 156 as a multiple of 120 and a remainder, and we carry on until the process terminates (as it must):

$$156 = 1 \times 120 + 36,$$
$$120 = 3 \times 36 + 12,$$
$$36 = 3 \times 12.$$

The greatest common divisor is the last remainder in the process, and thus $(432, 156) = 12$. The result (2) is implicit in the Euclidean algorithm; we need only work backwards as follows:

$$12 = 120 - 3 \times 36$$
$$= 120 - 3(156 - 120)$$
$$= 4 \times 120 - 3 \times 156$$
$$= 4(432 - 2 \times 156) - 3 \times 156$$
$$= 4 \times 432 - 11 \times 156.$$

What can we say about equation (1), in the light of these results? Evidently, if c in equation (1) happens to be equal to (a, b), then the equation has the special solution $x = k$, $y = l$. More generally, if c is some multiple of (a, b), say $c = q(a, b)$ (with q an integer), then equation (2) gives $a(qk) + b(ql) = q(a, b) = c$, so the original equation (1) has a special solution $x = qk$, $y = ql$. *Conversely*, suppose equation (1) has integer solutions x, y for given c, then c must be a multiple of (a, b).

Proof: (a, b) divides both a and b (by definition), and therefore it also divides the linear combination $ax + by$, which equals c. Thus we have shown that the necessary and sufficient condition for the solubility of equation (1) is that c must be a multiple of (a, b). We consider two examples:

(i) $7x + 11y = 13$. $(7, 11) = 1$, and $1 = 2 \times 11 - 3 \times 7$ (by inspection, or by Euclid's algorithm); so we have $7 \times (-3) + 11 \times 2 = 1$, therefore $7 \times (-39) + 11 \times 26 = 13$. Thus one solution is $x = -39$, $y = 26$. (*Note*: there are other solutions. *Exercise*: find *all* solutions!)

(ii) $3x + 6y = 22$. $(3, 6) = 3$, and 22 is not a multiple of 3; hence there are no integer solutions in this case.

A rather harder example of a Diophantine equation is $x^3 + y^3 = z^3 + w^3$. Although $x^3 + y^3 = z^3$ is insoluble (by Fermat's last theorem for $n = 3$), this equation has infinitely many solutions in integers (apart from trivial solutions such as $x = z$, $y = w$; $x = -y$, $z = -w$). The formulae were given by Euler; see Davenport (1970). There is a story connected with this equation, relating to the brilliant Indian mathematician S. Ramanujan who had a deep intuitive feeling for the integers. When his friend G. H. Hardy visted Ramanujan who was ill in hospital, he said he had come in taxi no. 1729 and this seemed rather a dull number, whereupon Ramanujan immediately said: 'No Hardy, it is very interesting – it is the smallest number expressible as the sum of two positive cubes in two different ways'. And in fact $1729 = 1^3 + 12^3 = 9^3 + 10^3$. 'Every positive integer was one of Ramanujan's personal friends' (Littlewood).

How about $x^4 + y^4 = z^4 + w^4$? Even Ramanujan could see no 'obvious' solutions in this case but he felt sure that any number expressible as the sum of fourth powers must be large. In fact it was known to Euler that

$$158^4 + 59^4 = 134^4 + 133^4 = 635\,318\,657.$$

3

Types of numbers

We have seen that there are many interesting and difficult problems connected with the simplest numbers, the integers, but you will already have come across other kinds of numbers: fractions, 'irrational' numbers like $\sqrt{2}$, and perhaps complex or 'imaginary' numbers. Let us now have a more general look at the concept of number, and the way it has evolved over the ages.

The simplest numbers are the 'natural numbers' $1, 2, 3, \ldots$, required for the process of *counting* objects. It was quite an abstract idea when primitive man realised at some stage, long ago, that 3 apples, 3 men, 3 women, 3 stones, etc., all have something in common: the (abstract) *number* 3. Let us briefly review some of the rules for calculation with natural numbers. (It is not necessary for our purpose to present a logically complete set of rules.)

The basic operations are addition and multiplication. They arise in a simple and natural way when we combine sets of objects. Thus, to calculate $2+3$, we start with $2 = $ ⊙⊙ objects and put next to them $3 = $ ⊙⊙⊙ objects; this gives us a collection of ⊙⊙⊙⊙⊙ $= 5$ objects (by counting). We arrive at the same number by starting with 3 objects and putting next to them 2 objects: ⊙⊙⊙⊙⊙ $=$ ⊙⊙⊙⊙⊙ . Thus we adopt as a general rule of addition $a+b = b+a$ for any two natural numbers (we call this the *commutative law of addition*). Multiplication arises in the form of repeated addition. 2×3 tells us to combine 2 collections of 3 objects. This contains altogether 6 objects, so $2 \times 3 = 6$. We can clearly regard the combination also as consisting of 3 collections of 2 objects (Fig. 3.1), so $3 \times 2 = 6$, and generally $a \times b = b \times a$ (this is the *commutative law of multiplication*). The link between addition and

Fig. 3.1

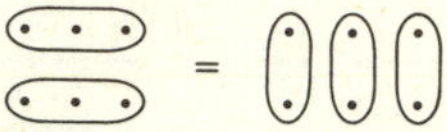

multiplication is important: $2 \times (3+4)$ appears as

We can regard this collection as made up of two 3s and two 4s:

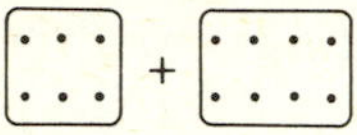

; thus $2 \times (3+4) = 2 \times 3 + 2 \times 4$, and generally $a(b+c) = ab + ac$. This is the *distributive law*.

These rules for the natural numbers, established by practical experience in counting, seem 'obvious' and hardly worth so much fuss. However, one of the basic ideas of mathematics is to abstract and generalise, and the arithmetic of the natural numbers leads us to study other 'objects' which satisfy the same general rules but with which we may no longer be able to count.

Note that, if a and b are any two natural numbers, then $a+b$ and $a \times b$ are also natural numbers, so that we can add and multiply natural numbers without any restriction. So far so good, but we find that our number system is incomplete as soon as we want to carry out *subtraction*. We can define subtraction by an equation: given two natural numbers a and b, we wish to find a number x such that $a+x = b$. Such an x exists as a natural number *only* if a is not greater than b (there is no natural number x such that $5 + x = 3$).

To resolve this difficulty we can do one of two things: *either* we remark only that, within the set of natural numbers, the equation $a + x = b$ has *no* solution when $a > b$ (and go away and study something else, as there is then very little to do in mathematics); *or* we can say that we want always to be able to solve the equation, and that we must therefore *extend* the number system beyond the natural numbers. To achieve our end we introduce the *negative integers* $-1, -2, -3, \ldots$; -2 for example is *defined* to be the number which satisfies $5 + x = 3$. With these new objects $a + x = b$ always has a solution, and in fact a unique solution denoted by $b - a$. The natural numbers 1, 2, 3, ... are then called *positive integers*, and the positive and negative integers, together with 0, form the set of all *integers*. This extension of the number system is of course not just an abstract game, but is needed for all sorts of applications where *gain* and *loss* are involved, for example stock market movements or thermometer readings. At this stage we introduce the important *geometric* representation of the integers as points on a line:

The positive integers then appear naturally as (evenly spaced) points to the *right* of zero, and the negative integers equally naturally as points to the *left* of 0. Evidently, if a, b are any two integers, then the point representing a on the number line is either to the right of the point representing b ($a > b$), or coincident with b ($a = b$), or to the left of b ($a < b$): we say that the set of integers is *ordered*.

Having introduced a new type of number, we must establish rules for calculating with these new objects. The rules for negative numbers will be familiar, and we will not spend time developing them systematically. But let us take a moment to consider why (for example) we say that $(-1) \times (-1) = +1$. If a child asks you this, what do you say? 'Because it is'? (That is, in a sense, the right answer!) First, we agree that (for example) $2 \times (-3) = -6$ (i.e. two losses of 3 is a loss of 6), etc. Next, we postulate that our general laws, established for addition and multiplication of natural numbers, should still hold for the set of all integers. In particular, we postulate the continued validity of the distributive law: $a(b + c) = ab + ac$. Then, choosing the values $a = -1$, $b = 1$, $c = -1$, we obtain $(-1) \times (1 - 1) = (-1) \times (1) + (-1) \times (-1)$, or $0 = -1 + (-1) \times (-1)$, which holds only if we have $(-1) \times (-1) = +1$. Thus the rule for multiplication of two negative numbers follows from the postulate that the distributive law is valid. Why should we postulate this law? There is no *logical* need to do so: it just happens to be the most 'sensible'.procedure for many purposes. We get the most 'useful' number system by preserving the rules established for the simplest numbers when we generalize: 'useful' both in the sense that the number system is needed for developing mathematics itself to higher levels, and also that it is most useful in applications of mathematics to practical problems of many kinds.

This attitude: 'let us adopt some rules (which we call 'axioms'), let us make sure as far as we can that they involve no logical contradictions, and then let us see what we can do with our rules', is a modern one in mathematics which emerged gradually in the nineteenth century. Before that, people tended to look for some underlying reality: they asked, do negative numbers (or complex numbers, or square roots) 'really' exist? So there were controversies over what was 'really' allowed in mathematics and what was not. For example, it was thought that Euclid's geometry was the only 'really' possible one, and only in the nineteenth century was it realised that non-Euclidean geometries with different axioms could be developed which had the same status as Euclid's from a logical point of view.

The mathematician nowadays no longer looks for such an underlying external reality. This philosophical modesty is richly rewarding, since it has released mathematicians to develop new mathematical structures much more freely and to explore their consequences. We may still ask: why then do we do mathematics, and what's the use? There is no simple answer, and the question is worth debating; philosophical arguments about the nature and purpose of mathematics have by no means died out!† Most mathematicians would try to answer the question at two levels, not necessarily with equal emphasis on both. They would firstly stress the use and importance of mathematical ideas within mathematics itself (thus without the real numbers, see p. 37, most of the interesting branches of mathematics could not be developed); and they would secondly point to their use and importance in applications (thus complex numbers are needed in applied mathematics in the analysis of vibrations; non-Euclidean geometry is essential for the modern theory of gravitation; group theory is important for quantum

---

† And it must be admitted that arguments about what is 'really' allowed' in mathematics have also continued, but at a more sophisticated level (see for example the reference to 'non-constructive proofs' in Chapter 11).

mechanics). It is worth remembering here that non-Euclidean geometry and group theory are branches of mathematics which were developed originally purely for their mathematical interest and whose important role in applications only became clear many years after their discovery. Thus one should not be too dogmatic in statements about which parts of mathematics are most 'relevant to the real world'.

The next generalisation of the number system arises when we want to do *division*. Given two integers a and b we wish to find x such that $ax = b$. Again, this equation has no solution among the integers except in the special case when b is a multiple of a (thus there is no integer x such that $3x = 5$). We deal with this problem by introducing the *rational numbers*. These are written in the form of fractions b/a, where a, b are integers. We exclude the case $a = 0$. (Division by zero must be excluded, since zero times any number is always zero, so there is no x such that $0 \times x = b$ if $b \neq 0$.) Such numbers must have originated very early on to deal with practical problems of subdivision such as sharing 5 sheep among 3 people; how much sheep for each person? A number of comments should be made. (1) The representation of rational numbers as fractions is not unique, since we want (for example) $2b/2a$ to be the same number as b/a (if you share 10 sheep among 6 people every person gets the same amount as when you share 5 among 3). We can obtain a unique representation by stipulating that a is positive and that a, b have no common factors; the fraction b/a is then 'in its lowest terms'. (2) A *field* in mathematics is essentially a collection of 'numbers' such that the sum, difference, product and quotient of any two numbers in the collection (excluding division by 0) are also in the collection. With the rationals we can perform these four operations, and the set of all rational numbers is the most familiar (but by no means the only) example of a field. (3) With the rationals we now have a wider class of numbers which includes the integers as a subset (they are rationals of the form $b/1$). (4) The rationals are an ordered field and can again be regarded as points on a line. They give us a subdivision of the line which we can make as fine as we like. To represent (for example) all rationals of the form $b/1\,000\,000$ as points on the line, we divide the interval $(0, 1)$

into a million equal pieces; similarly for all other intervals (1, 2), (2, 3), . . . ; and the points of subdivision then correspond to fractions of the form $b/1\ 000\ 000$. We can make the denominator as large as we like (10^{100} or what you will) – so one would surely be inclined to think that one must in this way eventually catch *all* the points on the line; in other words, that every point on the line is described by *some* rational number. It is certainly true that rational numbers are sufficient for all practical purposes of measuring. What is more, the rational points are *dense* on the line; this means that we can never find an interval (a, b), however small, between any two given rational points which is entirely *free* from rational points. To verify this we merely observe that, if a and b are different rational numbers with $a < b$, the rational number $\frac{1}{2}(a + b)$ lies between a and b. (*Exercise*: demonstrate this.) The fact that any interval between rational points contains at least *one* other rational point immediately implies, remarkably enough, that any such interval must contain *infinitely many* rational points! For, if there were only a finite number, say m, we could mark them off as shown:

and then any interval between two adjacent points would be free of rational points; but we have just seen that this is impossible. Thus there *appear* to be no 'empty places' on the line.

It was one of the most remarkable discoveries in mathematics, made by the Pythagoreans about 2400 years ago, that this very natural conclusion is not correct: there *are* numbers which are not rational and which cannot be obtained by *any* subdivision of the line into an integral number of pieces. In fact 'most' numbers are not rational (in a sense to be discussed later on; see Chapter 11)! The discovery of irrational magnitudes may be regarded as the beginning of theoretical 'pure' mathematics. What exactly was the problem worrying the Greek mathematicians here? A rule for constructing a right angle (the 'carpenter's rule') had been known for a long time: take two arms, of lengths 3 and 4 units, and incline them so that the line joining the ends is of length 5 (Fig. 3.2). It is natural to try to find a common unit of measure for the sides of the even simpler triangle which has two *equal* sides inclined at a right angle. Suppose we divide these sides into 5 equal pieces; then the hypotenuse is found by measurement to contain just over 7 of these pieces. Try then dividing into 12 parts: the hypotenuse (to a much better approximation) will then contain 17 pieces, but the subdivision is still not exact. All efforts to find a 'common measure' for all the sides of such a triangle were fruitless, and eventually it was *proved* that no such common measure could ever be found. Thus the equal sides and the hypotenuse of the triangle are 'incommensurable'. Note that to the Greeks 'number' meant 'whole number' (they didn't even deal with fractional numbers, but always talked instead about *proportions* or *ratios* of two whole numbers): they did not conceive of the notion of an irrational quantity as a *number*, but thought of incommensurable *geometrical ratios*.

Given two lines of length 1, we construct a line, whose length x is not rational, as the hypotenuse of the right-angled triangle whose shorter sides are 1: the Pythagorean theorem then says $x^2 = 1^2 + 1^2 = 2$. Thus the problem is: find a number x such that $x \times x = 2$. (Note the difference from the *linear* equation $ax = b$ which defines the rationals.) The proof that no such rational number x exists can be given in many ways. The usual modern version, which is equivalent to a proof given by Aristotle, is as follows: suppose $x = p/q$, a rational number in its lowest terms (so that p, q are integers without common factor); then $p^2/q^2 = 2$, i.e. $p^2 = 2q^2$, hence p^2 is an even number, hence p is even, since the square of an odd number is always odd. So we can write $p = 2r$, where r is an integer, i.e. $4r^2 = p^2 = 2q^2$, so $q^2 = 2r^2$, so q^2 is even, hence q is even. Thus p and q are both even,

Fig. 3.2

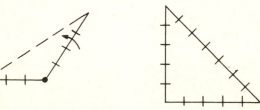

contradicting the assumption that p and q have no common factor. Denoting x by $\sqrt{2}$, we have shown that *no rational number is equal to* $\sqrt{2}$. The recognition that no common measure can be found for certain geometric magnitudes appears to be the first of many 'proofs of impossibility' which have played an important role in the development of mathematics.

It is not known for certain how incommensurability was first discovered, but it has been suggested that it could have arisen in connection with the method of subdivision of a regular pentagon, shown in Fig. 3.3, which was known to the Pythagoreans. If one starts with a regular pentagon $ABCDE$ and draws all five diagonals, then these diagonals intersect in points $A'B'C'D'E'$ which form another regular pentagon. The diagonals of this smaller pentagon form a still smaller regular pentagon, and so on. Clearly this process of subdivision never stops, and there is no 'smallest pentagon' whose side could serve as an ultimate unit of measure. It follows that the ratio of a side to a diagonal in a regular pentagon cannot be rational. (*Exercise*: show that the ratio is in fact $\frac{1}{2}(\sqrt{5}-1)$.)† This method of proof is called *infinite descent*: it was Fermat's favourite technique for demonstrating results in the theory of numbers.

Plato emphasised the importance of the discovery of incommensurability. He tells us that the mathematician Theodorus (born about 470 B.C.), a Pythagorean, had shown his pupils a proof that the side of a square of area 3 is incommensurable with the unit of measurement; similarly for squares of area 5, 6, 7, . . . up to 17 (excluding of course the squares 9 and 16).

To construct $\sqrt{2}$ as a length on the number axis we draw a right-angled triangle whose hypotenuse is $\sqrt{2}$ (Fig. 3.4) and use a compass to transfer the length $\sqrt{2}$ to the axis. This geometrical construction gives a point distinct from all rational points, so the set of rational points, although dense, does not cover the whole number line. This early and fundamental result shows that we must not trust 'first impressions' in mathematics. Care-

---

† To obtain this result, show first that the point D' divides AC in such a way that $AD'/D'C = D'C/AC$. Two thousand years later, this method of subdividing a line fascinated the painters of the Renaissance who called it the 'golden section'.

ful argument and subtle reasoning are needed and can lead to unexpected conclusions.

One can show, by extensions of the arguments used above, that other numbers formed by root extraction, such as $\sqrt{6}$, $\sqrt[3]{2}$ (these are solutions of the algebraic equations $x^2 = 6$, $x^3 = 2$), are also irrational. Further examples of irrational numbers are the familiar numbers π (the ratio of the circumference of a circle to its diameter) and e (the base of natural logarithms). These numbers can be defined in various equivalent ways, in particular by means of infinite series, but can they be obtained (like $\sqrt{2}$) as roots of algebraic equations with integer coefficients? We shall return to this interesting question later (see p. 67).

Fig. 3.3

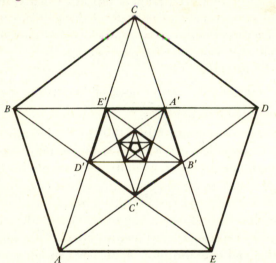

Fig. 3.4

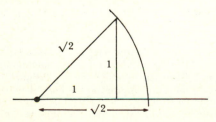

We now indicate another approach to irrational numbers which is not geometrical. This comes from the *decimal representation* of rational numbers. A decimal fraction represents a rational number if, and only if, it is periodic (or, as a special case, finite); e.g.

$$\tfrac{3}{4} = 0.75, \qquad \tfrac{2}{7} = 0.285\,714\,285\,714\,285\,714\ldots$$

To prove this result generally we look at the usual long division of B into A which gives the decimal form of A/B. The only possible remainders at each stage of the division are $1, 2, \ldots,$ $B - 1$ (i.e. there are $B - 1$ different possibilities); if any remainder is zero the process stops and we have a finite decimal. But this means that, after *at most B* divisions, one of the remainders *must* repeat, and thereafter the decimal representation repeats itself (since subsequent remainders repeat). Clearly, also, the 'repeating block' can have *at most $B - 1$* digits. Conversely, if the decimal representation of a number is periodic, then the number is rational. Consider for example

$$x = 1.818\,181\,8\ldots$$

We can convert this into a *finite* decimal by writing down

$$100x = 181.818\,181\,8\ldots$$

and subtracting: we obtain $99x = 180$, so $x = 180/99 = 20/11$. This reduction is evidently possible for any periodic decimal: if the repeating block contains m digits we multiply by 10^m and subtract. Thus we can say that any *non-periodic* infinite decimal must be an irrational number. It is not difficult to invent simple rules for writing down infinite non-periodic decimals which therefore represent irrational numbers; for example the following decimal containing an increasing number of zeros:

$$0.101\,001\,000\,100\,001\ldots$$

No simple rule of this kind, however, exists for constructing the decimal representation of commonly occurring irrational numbers such as $\sqrt{2}$.†

---

† There is a (not very serious) difficulty over uniqueness, arising from the fact that $0.999\ldots$ (with the 9s recurring indefinitely) is the same number as 1. Thus the infinite recurring decimal $0.359\,99\ldots$ represents the same number as the finite decimal 0.36. If we wish every number to have a *unique* decimal representation, we can agree to write every finite decimal as an infinite decimal with recurring 9s.

We can thus now regard a 'number' as *any* finite or infinite decimal, and this definition will include both rational and irrational numbers. This gives us the set of *real numbers*. Our discussion shows that, if we regard the integers as the basic 'building bricks' to be used in constructing all other kinds of number, then a rational number can always be defined in terms of two integers only, but the definition of an irrational number necessarily involves in some way an infinite set of integers. Infinite processes are thus inevitably involved in any proper theory of the real number system.

Such a precise logical theory was first given in the nineteenth century, and it accompanied the rigorous formulation of the calculus in the same period. In fact the simple definition of a real number as an infinite decimal, while adequate for many purposes, is not altogether satisfactory, since there is no special merit in the decimal system and the use of the base 10, and one would prefer a definition which is not linked to any particular base. And, more important, the definition should bring out the fundamental property of the real numbers (not possessed by the rationals) of being *complete*, i.e. (in a geometrical sense) the property of representing *all* the points on the number line. It is this completeness property, the fact that there are no 'gaps' left on the line, which is needed to give precise definitions of *continuity* and of a *continuous function*, the key concepts needed to put the calculus on a rigorous basis. One way of formulating the completeness property precisely is by way of the property of the *least upper bound*: we require that, if a collection of elements of our set has an upper bound, then it must have a *least* upper bound belonging to the set. This property is not possessed by the rationals: the collection of all rationals whose squares are less than 2 has an upper bound (they are all less than, say, 3), but no rational number is a least upper bound (this is because we can find rational numbers arbitrarily close to the irrational number $\sqrt{2}$ which does not itself belong to the set). These observations suggest an alternative way of defining the real numbers. We use the fact that, while the real number $\sqrt{2}$ is itself not rational, its properties are completely determined by specification of the (infinite) set of *all* rationals less than $\sqrt{2}$. But this amounts to saying that the real number can be *defined* as a *set* of rational numbers (which will have as one of its properties that it contains no greatest element). This is essentially the approach to the construction of the real numbers adopted by Dedekind (in 1872); it is a modern version of ideas which go back to the time of Eudoxus (see p. 44). There are other, equivalent, ways of constructing the set of real numbers; the technical details of the construction are matters for a university course in analysis. The essential properties of the set are that it is a *field* (see p. 31) which is both *ordered* (see p. 29) and also *complete*, in the sense just discussed. It can be shown that these properties make the real numbers *unique*, in the sense that there is only one complete ordered field, or, more precisely, that any two fields which are both complete and ordered have the same properties and are, for mathematical purposes, identical.

We now introduce briefly a further extension of the number concept, and we can then in the next chapter review the historical development of the various generalisations we have met.

We have seen that, to solve a quadratic equation like $x^2 = 2$, we need irrational numbers, and that the rational and irrational numbers make up the real numbers. These numbers are ordered and correspond to points on a line. Now, since the square of a real number, whether positive or negative, is always positive, there is no real number x such that $x^2 = -1$. Thus real numbers are not sufficient to solve a quadratic equation as simple as $x^2 + 1 = 0$ (and many others). So we must generalise the number concept again by introducing the symbol i (not a real number!) with the property $i^2 = -1$ (the 'imaginary unit'). With this we form *complex numbers* like $2 + 3i$, $-5i$, ..., and generally $a + bi$, where a, b are real numbers; a is the 'real' and b the 'imaginary' part of the complex number $a + bi$. We assume that the usual laws for addition and multiplication hold for complex numbers as for reals. We can then calculate with complex numbers as with real numbers, always replacing i^2, whenever it occurs, by -1. So, for example, the product $(2 + 3i)(1 + 4i)$ is the complex number $-10 + 11i$, because

$$(2 + 3i)(1 + 4i) = 2 + 8i + 3i + 12i^2$$

$$= (2 - 12) + (8 + 3)i = -10 + 11i.$$

We can easily show that the processes of addition, multiplication, subtraction and division of two complex numbers always lead to complex numbers of the form $a + bi$ (excluding division by 0), so the complex numbers form a *field*. This includes the field of real numbers as a subfield: real numbers are equivalent to complex numbers of the form $a + 0i$.

The equation $x^2 + 1 = 0$ now has the two solutions $x = i$ and $x = -i$. Of course we have gained much more: we can state, for example, that every quadratic equation $ax^2 + bx + c = 0$ (a, b, c real, $a \neq 0$) has precisely two roots, given by the well-known formula

$$x = \frac{-b \pm \sqrt{(b^2 - 4ac)}}{2a}.$$

The roots are real if $b^2 - 4ac \geqslant 0$, and complex if $b^2 - 4ac < 0$. In fact complex numbers do very much more than just this; they play a fundamental role in many branches of pure and applied mathematics.

There is no order relation among complex numbers of the kind that exists for real numbers, and we cannot represent them as points on a line. Nevertheless, a simple (and very important) geometric interpretation can be given; it was introduced in the early nineteenth century. This represents $x + yi$ as the point in a *plane* which has rectangular coordinates x, y. In this way any complex number is uniquely associated with a point in the 'complex plane' or 'Argand diagram', Fig. 3.5. Addition and multiplication of complex numbers also have simple geometric interpretations. For addition, we regard $x + yi$ as a displacement

Fig. 3.6

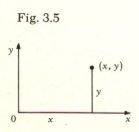
Fig. 3.5

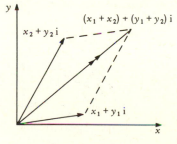

(a two-dimensional *vector*) from $(0, 0)$ to (x, y); then the sum

$$(x_1 + y_1 i) + (x_2 + y_2 i) = (x_1 + x_2) + (y_1 + y_2)i$$

is the point in the complex plane obtained by the vector (parallelogram) addition law (Fig. 3.6). This geometric interpretation shows that complex numbers are just as 'real' as 'real numbers'. We need only think in two dimensions instead of one. The names 'real' and 'imaginary' are historical, and remind us of earlier days when the 'square root of minus one' was regarded as something 'fictitious' or 'unreal', in fact 'imaginary'!

The geometric interpretation of multiplication by a complex number is also interesting. In particular, multiplication of $x + yi$ by a complex number of the form $\cos \theta + i \sin \theta$ is equivalent to a *rotation* of the vector represented by $x + yi$ through an anticlockwise angle θ in the complex plane. To verify this we use *polar coordinates* in the plane (Fig. 3.7), and write

$$x + yi = r \cos \alpha + i \, r \sin \alpha.$$

Then

$$(x + yi)(\cos \theta + i \sin \theta)$$
$$= r \, (\cos \alpha \, \cos \theta - \sin \alpha \, \sin \theta)$$
$$+ ir \, (\cos \alpha \, \sin \theta + \sin \alpha \, \cos \theta)$$
$$= r \cos (\alpha + \theta) + ir \sin (\alpha + \theta),$$

and this is a vector of the same length r as $x + yi$, making an angle $\alpha + \theta$ with the x-axis. In particular, taking $\theta = \frac{1}{2}\pi$ (a right angle), we have $\cos \frac{1}{2}\pi + i \sin \frac{1}{2}\pi = i$, and we see that multiplication by i is equivalent to rotation through a right angle. What, then, about the product $i^2 = i.i$? Rotating the vector i through a right angle again, we get to the point -1 in the complex plane (Fig. 3.8).

Fig. 3.7   Fig. 3.8

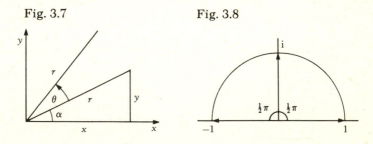

Thus we have a simple geometric interpretation of the fundamental rule $i^2 = -1$. The complex number $\cos \theta + i \sin \theta$ is a vector of unit length and corresponds to a point on the *unit circle* (with centre at the origin). It describes a rotation through an angle θ about the origin. Multiplication of two such complex numbers amounts to combining two rotations through θ_1 and θ_2, and gives a rotation through $\theta_1 + \theta_2$. This is a 'group operation', and the set of these complex numbers forms the *group of rotations* about the origin in two dimensions.

4

Historical development of the number concept I

The idea that we can freely create number systems to suit our requirements is a modern one, originating in the early nineteenth century. Let us now go back and review the situation in earlier periods.

We know that the Babylonians (about 2000 B.C.) had a well-developed system of mathematics which allowed them to solve quite tricky problems. It seems that they were not interested in general theory, only in computing answers (according to specified rules) to specific questions. They could deal with problems which gave rise to quadratic equations. For example: 'An area, consisting of the sum of two squares, is 1000. The side of one square is $\frac{2}{3}$ of the side of the other square, diminished by 10. What are the sides of the squares?' This leads to the equations $x^2 + y^2 = 1000$ and $y = \frac{2}{3}x - 10$, whence $13x^2 - 120x - 8100 = 0$, with positive root $x = 30$. The other (negative) root $x = -270/13$ was simply ignored. Were the Babylonians aware of the 'existence' of this second root? Probably not: they clearly wanted only to solve their practical problem of measurement and were happy to find a unique positive answer. Nowadays mathematicians are more interested in general properties such as the fact that quadratic equations always possess exactly two roots.

As we saw in Chapter 3 irrational magnitudes were discovered by the Pythagoreans. These people, a religious, scientific and philosophical brotherhood in Greece who lived in the period 600–400 B.C., were probably the first to recognise that mathematics deals with abstractions as distinct from physical objects or pictures. They dealt with whole numbers and their ratios only ('commensurable' ratios) and they assumed that all phenomena (in particular geometrical magnitudes) must be describable in terms of whole numbers. So the discovery that the two sides of a right-angled triangle can be incommensurable must have produced a serious crisis in Greek mathematics. (Legend ascribes the discovery to a certain Hippasus and says that he was thrown overboard as a punishment for shattering fundamental beliefs.) As the Pythagoreans did not accept irrational numbers the identification of number with geometry was destroyed.

Just why was this discovery such a serious matter? It demolished most of the Greek geometry of the time. Consider for example the theorem that for two triangles of the same height the areas are in the ratio of their bases, area A : area $B = a : b$ (Fig. 4.1). Let us assume that we have proved that the areas are equal

Fig. 4.1

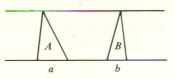

when the bases are equal ($a = b$). We can then easily prove our more general result when a and b are commensurable. For example let $a : b = 3 : 2$, then $2a = 3b$, so we can construct *equal* triangles by comparing 2 As and 3 Bs (Fig. 4.2). These have equal bases and therefore equal areas, so $2A = 3B$, i.e. $A:B = 3:2$. For incommensurable lengths all such proofs (i.e. all the theory of *proportion* and *similarity*) fail, and it is not clear how we can compare two areas at all in such cases. Here the problem of *continuity* appeared. Number to the Greeks is a discontinuous concept, but for geometry we need to deal with continuously variable magnitudes. The general problem of

Fig. 4.2

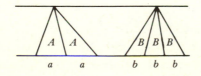

calculating areas and volumes of all kinds of figures (particularly those with curved boundaries) led on to the *calculus*, for
which one needs properly formulated concepts of *continuity*
and *limits*. We deal with these questions nowadays by starting
with a generalised concept of number. Our modern ideas had
their roots in the Greek approach to the problem of comparing
incommensurable magnitudes. They invented the following
special rule for dealing with magnitudes which have incommensurable ratios: suppose p, q are any two whole numbers and
we want to compare a and b with A and B. Since it may not be
possible to choose p and q such that $qa = pb$, we distinguish
three cases: either $qa < pb$, or $qa = pb$, or $qa > pb$. Then if,
whenever $qa < pb$, we have also $qA < pB$; *whenever* $qa = pb$ we
have $qA = pB$; and *whenever* $qa > pb$ we have $qA > pB$, we say
that $a : b = A : B$. This postulate forms the basis of Eudoxus's
theory of proportions, given by Euclid. It contains the case of
commensurable magnitudes (the middle possibility) as a special case, and it allows us once again to compare (for example)
the areas of triangles when the bases are incommensurable.
Consider q triangles with base a and p with base b, then we
shall have *one* of the cases $qa < pb$, $qa = pb$ or $qa > pb$; suppose
the first. But suppose two triangles with areas U, V have the
same height and have bases u, v with $u < v$, then also $U < V$ (see
Fig. 4.3). It follows that, whenever $qa < pb$, we have also $qA <
pB$, and similarly in the other two cases. Thus, in accordance
with Eudoxus's postulate, we can conclude that $a : b = A : B$.
The nineteenth-century theory of irrational numbers follows
this mode of thought quite closely, translated into modern
language. Eudoxus's theory superseded the arithmetical theory

Fig. 4.3

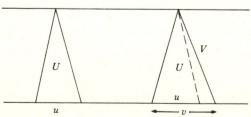

of the Pythagoreans which applied to commensurable quantities only. It cast algebraic reasoning into geometric form ($\sqrt{A}$ is
the side of a square of area A, and so on), and it is presented by
Euclid in strict axiomatic form.

Geometrical thinking thus became the basis for almost all
rigorous mathematics for about 2000 years. We still call x^2
'x squared', x^3 'x cubed', because these symbols had only
geometrical meaning for the Greeks. The intellectuals of classical Greece were philosophers with little interest in the
requirements of commerce and trade. Computation played little
part in their science which tended to be qualitative in character.
Euclid's *Elements*, which present an organised account of the
mathematical work of classical Greece, consist of geometry and
theory of numbers: the 'purest' branches of the subject. In the
later stages of the Alexandrian period educated men became
more interested again in practical affairs and the emphasis
shifted to quantitative knowledge and the development of
arithmetic and algebra. In particular Archimedes (287–212 B.C.)
was not only a highly original pure mathematician, but also a
master of calculating technique and a prolific inventor of
mechanical devices. In Chapter 8 we shall discuss his contributions to the early calculus.

Throughout the Hindu–Arab–Medieval period and well into
the nineteenth century people worried about the meaning of
negative and complex numbers. Such entities kept on making an
appearance in calculations, for example as unwelcome 'extra'
roots in the solution of quadratic equations, but what did they
mean? Diophantos, already in the third century A.D., was prepared to multiply together algebraic expressions like $(x-1)$ and
$(x-2)$, and to say that one must replace $(-1) \times (-2)$ by $+2$ in the
expansion. But he took it for granted that x had always to be
such that $x-1$ and $x-2$ were both positive; there was at that
time no suggestion that negative numbers could have any
meaning standing by themselves. Also quadratic equations
could have only positive roots. The assumption that mathematics dealt with positive numbers only was commonly held until
A.D. 1600 and later. The Indian mathematician Bhaskara (about
A.D. 1150) went so far as to give the roots of $x^2 - 45x = 250$ as

$x = 50$ and $x = -5$, but said: 'the second value is in this case not to be taken, for it is inadequate; people do not approve of negative roots'.

During the sixteenth and seventeenth centuries mathematicians began to operate with complex numbers, but without any real understanding of their nature. Thus Cardan (in 1545) considers the problem of dividing 10 into two parts whose product is 40. He obtains the answer $5 + \surd(-15)$, $5 - \surd(-15)$ and says: 'putting aside the mental tortures involved, multiply these numbers together; the product is $25 - (-15) = 40$. So progresses arithmetic subtlety the end of which, as is said, is as refined as it is useless'. (See also Chapter 5.) Descartes, who laid the foundation of coordinate geometry in his *Géométrie* (in 1637), also rejected complex roots and coined the term 'imaginary'. Even Newton did not regard complex roots as significant, presumably because in his day they lacked physical meaning. Leibniz wrote: 'The Divine Spirit found a sublime outlet in that wonder of analysis, that portent of the ideal world, that amphibian between being and not-being, which we call the imaginary root of negative unity'. (Should we regret that such colourful language is no longer used by mathematicians?)

By 1700 all the familiar members of our number system were known, but opposition to the newer types of numbers continued throughout the eighteenth century, and it was considered mathematically respectable to formulate problems so as to avoid them if possible. This could not always be done. Thus, for example, there was confusion in the early part of the century about the correct values of logarithms of negative numbers.† Johann Bernoulli argued as follows: since $(-x)^2 = x^2$ it follows that $2 \log(-x) = 2 \log x$, hence $\log(-x) = \log x$, and in particular $\log(-1) = \log 1 = 0$. Leibniz disagreed. He pointed out that, if

---

† Logarithms of negative and complex numbers made their appearance in the evaluation of simple real integrals by 'partial fractions'. Thus

$$\int \frac{dx}{x^2 + 1} = \frac{1}{2i} \int \left(\frac{1}{x - i} - \frac{1}{x + i} \right) dx = \frac{1}{2i} \log \frac{x - i}{x + i}.$$

Without understanding logarithms of complex numbers one could not show that this strange-looking result was the same as $\tan^{-1} x$.

$y = \log x$, then

$$x = e^y = 1 + y + \frac{y^2}{2!} + \cdots$$

If $\log(-1) = 0$, this equation would have to hold when $x = -1$ and $y = 0$. This is clearly not the case, and Leibniz maintained that logarithms of negative numbers had to be imaginary. Such a fundamental disagreement between two leading mathematicians of the period was disturbing. The matter was clearly resolved by Euler (about 1747) with the aid of the fundamental formula $e^{i\theta} = \cos\theta + i\sin\theta$. When $\theta = \pi$ this gives $e^{i\pi} = -1$, so $\log(-1)$ has the purely imaginary value $i\pi$: Leibniz was right! Euler showed furthermore that the disagreements about the correct value for the logarithm arose from its *multivalued* character. In modern form his argument amounts to saying that $e^y = e^{y+2in\pi}$ for all integral n and therefore, if $\log x = y$, then $y + 2in\pi$ is also a natural logarithm of x for every n. (*Exercise*: find the error in Johann Bernoulli's argument.) Although Euler's reasoning was remarkably clear, the idea of a 'multivalued function' was one that his contemporaries found hard to accept. The mathematicians of the eighteenth century were concerned primarily to *use* mathematics, following the creation of the calculus as a powerful tool for scientific problems. In their quest for results they were usually content to apply intuitively known rules of operation, but – as the example of the logarithm shows – reliance on crude intuition could lead to confusion and contradictions, and towards the end of the eighteenth century it became increasingly apparent that better logical foundations were needed for the calculus and for the number systems used in mathematics.

Mathematicians began to feel much happier about complex numbers when their geometric interpretation became familiar in the years after 1800. Important new mathematical concepts do not usually have a single discoverer, and the *idea* of complex numbers as points in a plane was already 'in the air' in Euler's time.[†] The main credit for the earliest clear recognition of the

---

[†] Indeed the idea occurs in rudimentary form already in John Wallis's *Treatise of Algebra* (1685). It was however ignored by his contemporaries.

correspondence between complex numbers and points in a plane is nowadays given to three men: C. Wessel, a Norwegian-born surveyor, who in 1797 emphasised the picture of a complex number as a two-dimensional vector (his paper, published in the transactions of the Danish Academy, remained unnoticed for 100 years); J. R. Argand, a Swiss self-taught bookkeeper, whose publication (in 1806) emphasised the interpretation of multiplication by i as a rotation through 90°; and Gauss, who indeed anticipated many of the most important nineteenth-century mathematical discoveries (for example elliptic functions and non-Euclidean geometry) without publishing them, and who seems to have been in full possession of the geometric theory of complex numbers by 1815. Nevertheless, as late as 1831 the distinguished mathematician Augustus de Morgan (one of the originators of modern mathematical logic) could say, in his book *On the Study and Difficulties of Mathematics*, 'The imaginary expression $\sqrt{(-a)}$ and the negative expression $-b$ have this resemblance, that either of them occurring as the solution of a problem indicates some inconsistency or absurdity. As far as real meaning is concerned, both are equally imaginary, since $0-a$ is as inconceivable as $\sqrt{(-a)}$.' Even some twentieth-century textbooks on trigonometry were still trying to avoid the use of $\sqrt{(-1)}$.

5

The cubic equation

One particular development which occurred in the sixteenth century is worth discussing in more detail. Throughout the period from about 1200, the time of Fibonacci, to about 1500, the revival of mathematics in Western Europe had consisted of a rediscovery, by way of translations of the ancient works, of the mathematics of the Greeks and Arabs. It was an exciting surprise when Italian mathematicians discovered, early in the sixteenth century, a new mathematical theory which the Ancients and the Arabs had missed. This was the *general algebraic solution of the cubic equation,* and it has a curious history.

The rich commercial life of the Italian cities of the fifteenth century, and their traffic with the Orient, led to the development of improved methods of calculation for practical purposes, needed as aids for book-keeping and navigation. Several of the great painters and architects of the early Renaissance were good mathematicians: their intense interest in the use of *perspective* for the plane representation of three-dimensional objects led them to study the laws of solid geometry. The state of knowledge in the late fifteenth century in arithmetic, algebra and trigonometry is summed up in Pacioli's *Summa di Arithmetica*, printed in 1494 (it was one of the first mathematical books to appear in print). Pacioli says at the end of this book that the solution of cubic equations like $x^3 + mx = n$ is 'as impossible at the present state of science as squaring the circle'.

Here began the work of the mathematicians of the University of Bologna. This university was, around 1500, one of the largest and most famous in Europe. Let us discuss what is involved in the problem of the cubic equation.

We all know how to solve a *quadratic equation*: $x^2 + bx + c = 0$. We suppose for the present, with the Middle Ages in mind, that b and c are positive numbers, and we look for real roots. How is the solution expressed by the standard 'quadratic formula' obtained? We first note that we can *remove* the term bx which is linear in x by making the substitution $x = y - \frac{1}{2}b$: then $x^2 = y^2 - by + \frac{1}{4}b^2$, and the equation for the new unknown y is

$$y^2 - by + \frac{1}{4}b^2 + b(y - \frac{1}{2}b) + c = 0, \qquad \text{i.e. } y^2 = \frac{1}{4}b^2 - c.$$

This contains no term linear in y, and we obtain y by simply taking the square roots of the number on the right-hand side (assuming they exist!):

$$y = \pm\sqrt{(\tfrac{1}{4}b^2 - c)}, \qquad \text{giving } x = -\tfrac{1}{2}b \pm \sqrt{(\tfrac{1}{4}b^2 - c)}.$$

Taking a modern view, we note that this process always gives two (real or complex) roots for any two (real or complex) numbers b and c (the roots may be coincident). Historically this method of solution had been known from the time of the Babylonians and had been used to obtain the (real) roots of interest in specific problems. Whenever $\frac{1}{4}b^2 \geqslant c$ the formula gives two real roots without any trouble (provided that negative roots are allowed); in other cases it would have been said that there are no roots.

The mathematicians of Bologna showed that a similar, but more complicated, formula can be derived for the general cubic equation $x^3 + ax^2 + bx + c = 0$. (Note that this was not an achievement of any *practical* significance. People knew quite well at that time how to obtain *approximate* numerical solutions of cubic equations which were sufficiently accurate for all practical purposes. The problem, a purely theoretical one, was to derive an exact algebraic *formula* expressing the roots of the cubic equation in terms of the coefficients a, b, c appearing in the equation.) The method of solution goes as follows: we first remove the quadratic term ax^2 by putting $x = y - \frac{1}{3}a$ (the coefficient of y^2 in the new equation is then $-3(\frac{1}{3}a) + a = 0$). The new equation is thus of the form $y^3 + py + q = 0$, where the coefficients p and q are easily expressible in terms of the original coefficients a, b and c. Now put $y = u + v$, then

$$y^3 = u^3 + 3u^2v + 3uv^2 + v^3 = u^3 + v^3 + 3uv(u + v)$$
$$= u^3 + v^3 + 3uvy.$$

Hence the equation becomes

$$u^3 + v^3 + (3uv + p)y + q = 0.$$

Now *choose* u and v to satisfy

$$3uv + p = 0.$$

Then we have

$$u^3 + v^3 = -q, \qquad u^3v^3 = -p^3/27.$$

These are two simultaneous equations for the two unknowns u^3 and v^3. If we eliminate one of these, we find that the other satisfies the *quadratic equation*

$$t^2 + qt - \frac{p^3}{27} = 0.$$

The two roots of this equation are u^3, v^3 and thus, using the quadratic formula to write down these roots, we have

$$u^3 = -\tfrac{1}{2}q + \sqrt{\left(\frac{q^2}{4} + \frac{p^3}{27}\right)}, \qquad v^3 = -\tfrac{1}{2}q - \sqrt{\left(\frac{q^2}{4} + \frac{p^3}{27}\right)},$$

and finally

$$y = u + v = \left[-\tfrac{1}{2}q + \sqrt{\left(\frac{q^2}{4} + \frac{p^3}{27}\right)}\right]^{1/3}$$

$$+ \left[-\tfrac{1}{2}q - \sqrt{\left(\frac{q^2}{4} + \frac{p^3}{27}\right)}\right]^{1/3}.$$

Thus we have derived a formula for the roots of the cubic equation. Let us leave the mathematics for the moment to have a look at the tortuous history of this discovery.

The solution is called 'Cardan's solution', but the first man to find it was Scipione del Ferro (about 1500). He was a professor of mathematics at Bologna who told a pupil of his discovery but did not publish the result. Nowadays a mathematician with an important new result is only too anxious to see it in print, but at that time it was usual to keep one's discoveries secret, so as to secure an advantage over one's rivals by proposing problems for solution which were beyond their reach. Hence disputes over priority were common. It is believed that the result was discovered independently (about 1530) by Nicolo of Brescia (called Tartaglia, the stammerer) who was a powerful self-taught mathematician. He challenged del Ferro's pupil to a public discussion (1535), in which each contestant gave thirty problems to the other. Such contests aroused great interest, like football matches today, and Tartaglia, who won easily, became famous. He kept his method for the cubic equation secret, saying he would publish it eventually. But a certain Cardano from Milan ('a singular mixture of genius, folly, self-conceit and mysticism', Cajori 1924, p. 134) persuaded Tartaglia to reveal the method, giving a solemn promise of secrecy. He then proceeded to publish the solution (in 1545) in his *Ars Magna* (giving, it is true, the credit for the discovery to Tartaglia). This betrayal shattered Tartaglia (to quote Cajori again: 'his most cherished hope, of giving to the world an immortal work which should be the monument of his deep learning and power for original research, was suddenly destroyed; for the crown intended for his work had been snatched away'). Bitter disputes, and hectic problem-solving contests, ensued between Tartaglia and Cardan (and the latter's pupil Ferrari). Finally Tartaglia started writing his great work after all, but he died before he reached the topic of cubic equations. Thus the method became known as Cardan's solution of the cubic, although Cardan was not the first discoverer. The complete details of the discovery are not known, nor are they very important; and we need not feel too sorry for Tartaglia, who was himself guilty of publishing works without giving due credit.

Our algebraic method of solution looks simple enough, and you may wonder why the discovery should have caused so much excitement. One must remember that for Cardan and his contemporaries the task was long and difficult: where we use a simple binomial expansion he used geometrical theorems which were much harder to handle, and he also had to consider separately the cases $x^3 = mx + n$, $x^3 + mx + n = 0$, $x^3 + n = mx$, $x^3 + mx = n$, because only positive numbers were allowed to appear in the equations, and a separate geometrical demonstration was needed for each type. Our account of the mathematics is given in modern notation, and it must not be thought that Cardan wrote down symbols such as are used here. Where we write $x^3 + mx = n$, Cardan would use geometrical language, with specific numerical coefficients, and say 'let the cube and five times the side be equal to eighteen' (he regarded the equation $x^3 + 5x = 18$ as typical of all those 'having a cube and a multiple of a side equal to a number'). Advances in mathematics often followed rapidly upon the invention of improved notation; in the calculus this was the achievement of Leibniz (see p. 123), and in elementary algebra much of the transition towards our modern forms took place in the hundred years or so between Cardan and Descartes.

How does Cardan's method of solution work in practice? Consider say the equation $x^3 - 2x^2 + 2x - 1 = 0$. Putting $x = y + \frac{2}{3}$ this becomes $y^3 + \frac{2}{3}y - \frac{7}{27} = 0$, so $p = \frac{2}{3}$, $q = -\frac{7}{27}$, and

$$\frac{q^2}{4} + \frac{p^3}{27} = \frac{49 + 32}{4.27^2} = \frac{81}{54^2},$$

thus

$$\pm \sqrt{\left(\frac{q^2}{4} + \frac{p^3}{27}\right)} = \pm \frac{9}{54},$$

and the Cardan formula gives

$$y = \left(\frac{7}{54} + \frac{9}{54}\right)^{1/3} + \left(\frac{7}{54} - \frac{9}{54}\right)^{1/3} = \left(\frac{8}{27}\right)^{1/3} + \left(-\frac{1}{27}\right)^{1/3}$$

$$= \frac{2}{3} - \frac{1}{3} = \frac{1}{3}$$

(taking real cube roots!); hence $x = y + \frac{2}{3} = 1$.

In fact $x^3 - 2x^2 + 2x - 1 = (x - 1)(x^2 - x + 1)$, so we can see without using Cardan's formula that there is just one real root $x = 1$ (but of course the formula can be used also when there is no such simple factorisation). Also the graph of $x^3 - 2x^2 + 2x - 1$ (Fig. 5.1) shows that there is *one* real root in this case. However, the cubic equation $x^3 + ax^2 + bx + c = 0$ with real coefficients a, b, c may have either *one* or *three* real roots, and Cardan's formula leads to trouble when there are three real roots (this is the so-called 'irreducible' case). To see what happens we consider as an example the equation $x^3 - 2x^2 - x + 2 = 0$. The left-hand side factorises as $(x + 1)(x - 1)(x - 2)$, so the equation has the three real roots -1, 1, 2. What does Cardan's formula give us? Putting $x = y + \frac{2}{3}$ gives

$$y^3 - \frac{7}{3}y + \frac{20}{27} = 0, \quad \text{so} \quad p = -\frac{7}{3}, \quad q = \frac{20}{27},$$

and now

$$\frac{q^2}{4} + \frac{p^3}{27} = \frac{400 - 1372}{4.27^2} = -\frac{972}{54^2} = -\frac{1}{3},$$

and Cardan's formula tells us to take the square root of this number which is unfortunately negative! Thus, although the roots of the original cubic equation are known to be real, Cardan's method fails to produce the solution if we are not prepared to use complex numbers. The formula gives

$$y = \left[-\frac{10}{27} + \sqrt{\left(-\frac{1}{3}\right)} \right]^{1/3} + \left[-\frac{10}{27} - \sqrt{\left(-\frac{1}{3}\right)} \right]^{1/3}$$

$$= \left(-\frac{10}{27} + \frac{i}{\sqrt{3}} \right)^{1/3} + \left(-\frac{10}{27} - \frac{i}{\sqrt{3}} \right)^{1/3}.$$

How can this complicated-looking expression give a real result? In fact it does do so because, while the cube roots in this formula are all (individually) complex, their *sum* turns out to be real! Every complex number has exactly three cube roots; the cube

Fig. 5.1

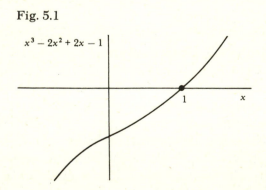

$x^3 - 2x^2 + 2x - 1$

roots in the expression for y turn out to be

$$u = \tfrac{1}{6}(-5+\sqrt{3}i), \qquad v = \tfrac{1}{6}(-5-\sqrt{3}i),$$
$$\tfrac{1}{6}(1-3\sqrt{3}i), \qquad \tfrac{1}{6}(1+3\sqrt{3}i),$$
$$\tfrac{1}{6}(4+2\sqrt{3}i), \qquad \tfrac{1}{6}(4-2\sqrt{3}i).$$

(You can check that this is correct by cubing these expressions and using $i^2 = -1$.) Why does this not give *nine* values of y ($y = u + v$)? We must combine u and v here to produce a real y (in general, u and v must satisfy $3uv + p = 0$, see p. 51); and this condition restricts us to the three values $y = -\tfrac{5}{3}$ or $\tfrac{1}{3}$ or $\tfrac{4}{3}$; therefore $x = y + \tfrac{2}{3} = -1$ or 1 or 2, as required.

This problem of complex numbers appears whenever the cubic equation has three real roots. Cardan noted the difficulty but did not know what to do about it. Some progress in recognising the nature of the problem was made by a later sixteenth-century Bologna mathematician, R. Bombelli (his *Algebra* appeared in 1572). He *guessed* that, if (say) the real root $y = \tfrac{1}{3}$ is to come from

$$\left(-\frac{10}{27} + \frac{i}{\sqrt{3}}\right)^{1/3} + \left(-\frac{10}{27} - \frac{i}{\sqrt{3}}\right)^{1/3},$$

then the cube roots are presumably of the form $\tfrac{1}{6} + \lambda\sqrt{(-1)}$ and $\tfrac{1}{6} - \lambda\sqrt{(-1)}$ (i.e. they are what we call 'conjugate complex numbers'); it is then easy to verify that λ must be $-\tfrac{1}{2}\sqrt{3}$. But this approach succeeds only if the answer is already known: if one tries to find the cube roots of complex numbers algebraically (i.e. given a and b, tries to find x and y such that $(x + iy)^3 = a + ib$), then one merely returns to the original cubic equation (that's why this is called the 'irreducible' case). It was only much later, about 1730, that it was clearly recognised (by Euler) that Cardan's solution always gives precisely three roots (possibly repeated); and it was only then that a systematic method of calculating the roots of complex numbers was given.

Another way of dealing with the irreducible case is to make use of the trigonometric identity $\cos^3 A - \tfrac{3}{4}\cos A - \tfrac{1}{4}\cos 3A = 0$. This can be identified, after a slight transformation, with a given cubic equation in the form $x^3 + 3px + q = 0$. Put $x = \cos A/m$, then the cubic equation becomes

$$\cos^3 A + 3m^2 p \cos A + m^3 q = 0.$$

So, in order to satisfy the equation, we put $3m^2 p = -\frac{3}{4}$ and $m^3 q = -\frac{1}{4} \cos 3A$. We can then calculate m and hence $\cos 3A$ in terms of the coefficients p and q. We now obtain the angle $3A$ from a table of cosines. We thus have A, and hence the root $\cos A/m$ of the original equation. There are two additional roots $\cos(A+120°)/m$, $\cos(A+240°)/m$, which correspond to the same value of $\cos 3A$. (*Exercise*: why are there precisely *three* roots?) This method is due to Francis Vieta who was the foremost French mathematician of the sixteenth century (he was a lawyer who did mathematics in his spare time). Vieta's approach avoids the use of complex numbers in the irreducible case.

Vieta in fact knew the general formulae which express $\cos nx$ and $\sin nx$ in powers of $\cos x$ and $\sin x$. These enabled him to solve a fearsome-looking problem, characteristic of the age, proposed as a challenge by the Belgian A. van Roomen (in 1593): solve the equation

$$x^{45} - 45x^{43} + 945x^{41} - 12300x^{39} + \cdots -3795x^3 + 45x = K.$$

Vieta, noting that the equation arises in the expansion of $\sin 45\theta$ in terms of $\sin \theta$, found all the positive roots and vindicated the honour of French mathematics.

It is curious that the first use of imaginary numbers was in the theory of *cubic* equations, and not in the theory of quadratic equations where we usually introduce them now. In the case of the cubic equation it was clear that real solutions existed, though in a strange form, and the appearance of imaginary quantities could not be avoided by saying that the solution did not exist. From this time on, complex numbers began to lose their mystery, though they were not fully accepted until the nineteenth century.

After the solution of the cubic equation it was natural to try to deal similarly with equations of order 4 (*quartic* or *biquadratic* equations). Cardan had himself considered some special cases, and the general solution was soon found by Cardan's pupil and secretary Ferrari. Let us see how it goes.

There are various equivalent methods: all of them reduce the problem to the solution of cubic and quadratic equations. One version proceeds as follows. Consider

$$x^4 + ax^3 + bx^2 + cx + d = 0;$$

add $(ex + f)^2$ to both sides, obtaining

$$x^4 + ax^3 + (b + e^2)x^2 + (c + 2ef)x + (d + f^2) = (ex + f)^2. \quad (*)$$

Choose e and f to get a perfect square on the left; suppose the left-hand side is $(x^2 + px + q)^2$. Squaring this and comparing, we have

$$2p = a, \quad p^2 + 2q = b + e^2, \quad 2pq = c + 2ef, \quad q^2 = d + f^2.$$

This fixes p. Rewrite the other equations as

$$e^2 = p^2 + 2q - b, \quad 4e^2f^2 = (2pq - c)^2, \quad f^2 = q^2 - d.$$

Substitute the first and third of these equations into the second:

$$(2pq - c)^2 = 4(p^2 + 2q - b)(q^2 - d),$$

or (using $p = \tfrac{1}{2}a$)

$$(aq - c)^2 = (a^2 + 8q - 4b)(q^2 - d).$$

This is a *cubic* equation for q and can be solved for q in terms of a, b, c, d. Any root will do; then e and f can be obtained from the above equations. (*) becomes

$$(x^2 + px + q)^2 = (ex + f)^2,$$

or

$$[x^2 + (p + e)x + (q + f)][x^2 + (p - e)x + (q - f)] = 0,$$

i.e. we have the two *quadratic* equations

$$x^2 + (p + e)x + (q + f) = 0,$$
$$x^2 + (p - e)x + (q - f) = 0,$$

whose roots are the four roots of the original equation.

As an example we use the method to solve

$$x^4 - 2x^2 + 8x - 3 = 0.$$

Here $a = 0$, $b = -2$, $c = 8$, $d = -3$. Thus $p = 0$, and the cubic equation for q is $64 = (8q + 8)(q^2 + 3)$, which reduces to $q^3 + q^2 + 3q - 5 = 0$, or $(q - 1)(q^2 + 2q + 5) = 0$. We choose as solution $q = 1$ (the only real solution!), and have

$$e^2 = 2 + 2 = 4, \quad e = 2; \quad f^2 = 1 + 3 = 4, \quad f = -2$$

(we must be careful here to choose the signs so that the equation $2pq = c + 2ef$ is satisfied). The two quadratic equations are

$$x^2 + 2x - 1 = 0, \qquad x^2 - 2x + 3 = 0,$$

with solutions

$$x = -1 \pm \sqrt{2}, \qquad 1 \pm \sqrt{2}i.$$

The details of this process are not very important; what *is* theoretically significant is that there are general methods, involving only operations in the field of complex numbers (addition, subtraction, multiplication and division), together with extraction of roots, which enable us to express the roots of any polynomial equation of degree up to 4 in terms of the coefficients in the equation. This, then, was in essence known before 1600.

6

Historical development of the number concept II

The algebraic solution of the cubic and quartic equations which we discussed in the last chapter suggests a number of questions, of general significance, which were not properly answered for another 200 years or more.

Firstly: we have seen that, within the field of real numbers, a simple equation like $x^2 + 1 = 0$ has no solution, but in the field $\mathbb{C}$ of complex numbers it has exactly two roots ($\pm i$). Consider then the *general* algebraic equation of degree n

$$x^n + a_1 x^{n-1} + a_2 x^{n-2} + \cdots + a_{n-1}x + a_n = 0, \tag{1}$$

where the as are any (real or complex) numbers. We ask: (i) does this equation always have a root (in the field $\mathbb{C}$)? And (ii), if *yes*, just how many roots are there? Once question (i) has been answered with 'yes', it is relatively easy to answer question (ii). For the cubic equation, as we have noted, it was Euler who first clearly recognised (1732) that there are always exactly three roots. For the general equation (1) the fundamental result is that the equation *always has a solution in* $\mathbb{C}$. From this it is easy to show, by successive applications of the fundamental result, that (1) can be *factorised* into an expression of the form

$$(x - c_1)(x - c_2) \cdots (x - c_n) = 0 \tag{2}$$

(the theory of factorisation of polynomials is very similar to the theory of factorisation of whole numbers); this result shows that equation (1) has exactly *n roots* in $\mathbb{C}$. This conclusion is remarkable. It means that, with the introduction of complex numbers, we have 'finished the job' as far as the solution of algebraic equations is concerned. The field $\mathbb{C}$ suffices to solve all algebraic equations and we do not have to invent any further kinds of numbers. We say that the field is *algebraically closed*.

The first more or less satisfactory proof of the fundamental result was given by Gauss in his doctoral thesis (in 1799). Gauss's great insight led him to formulate the problem in the 'correct' way which allowed a simple general result to be derived. Before his time, when complex numbers had not been fully accepted, people studied instead the factorisation of polynomials with real coefficients and found of course that these could not always be factorised, as in (2), into linear factors with real coefficients.†

Gauss was very fond of this theorem and later gave two further demonstrations. His first proof used geometrical arguments, not entirely rigorous by modern standards. But Gauss's proof was very original for its time, and helped to inaugurate a new approach to the question of *mathematical existence*. The Greeks had already recognised that the existence of mathematical entities must be established before one can try to prove theorems about them. Their test for existence was (geometrical) constructibility. In later work on equations, existence was established by obtaining a *formula* displaying the quantity in question: for example Cardan's formula actually exhibits the quantity which satisfies the cubic equation. From 1600 to 1800 people (very naturally) tried to obtain analogous formulae for the solutions of algebraic equations of degree greater than 4. It was assumed without question that, if a solution exists, then it must be displayable as a formula; and it was very puzzling that, in spite of all the efforts over such a long period, no such formula could be found. Gauss realised that, in order to show that a root *exists*, we do not in fact have to be able to give a formula for computing it. He argued (more or less) as

---

† They found, instead, that a factorisation into *linear* and *quadratic* factors was always possible. This result follows at once from (2) and the fact that, if the as in (1) are real and $c_1 = \alpha + i\beta$ is a root of (1), then $\bar{c}_1 = \alpha - i\beta$ is also a root. Two linear factors thus combine into a quadratic factor with real coefficients:

$$[x - (\alpha + i\beta)][x - (\alpha - i\beta)] = x^2 - 2\alpha x + (\alpha^2 + \beta^2).$$

Examples:

$$x^4 + 1 = (x^2 + x\sqrt{2} + 1)(x^2 - x\sqrt{2} + 1),$$
$$x^4 + x^2 + 1 = (x^2 + x + 1)(x^2 - x + 1).$$

follows: for complex z $(z = x + iy)$ equation (1) is

$$(x + iy)^n + a_1(x + iy)^{n-1} + \cdots + a_{n-1}(x + iy) + a_n = 0. \quad (3)$$

Write each a_i as $\alpha_i + i\beta_i$, multiply out and collect real and imaginary parts. This leads to an equation of the form $u(x, y) + iv(x, y) = 0$, where $u(x, y)$, $v(x, y)$ are polynomials in the real variables x, y with real coefficients; and this implies that $u(x, y) = 0$ and $v(x, y) = 0$. These are the equations of two *curves* in the x, y-plane. Gauss, by studying the general form of these curves, was able to show that they must always *intersect* at least once. To do this he looked at the form of the curves *far from the origin* (where only the highest powers in the polynomials matter); he found that the curve $u(x, y) = 0$ has a branch with asymptotic directions defined by lines making angles $\theta = \pi/2n$ and $\theta = 3\pi/2n$ with the x-axis, and the curve $v(x, y) = 0$ has a branch with asymptotic directions defined by $\theta = 0$ and $2\pi/2n$ (Fig. 6.1). Since the curves are continuous it is clear from the figure that they must intersect *somewhere*, and the point of intersection defines a complex number $x + iy$ which satisfies equation (3).

As an example consider $z^3 - 2i = 0$. Here $n = 3$, $u(x, y) = x^3 - 3xy^2$, $v(x, y) = 3x^2y - y^3 - 2$ and $\pi/2n = \pi/6$. In the positive quadrant the curve $u = 0$ consists of the y-axis and the straight line $y = x/\sqrt{3}$ which makes an angle $\pi/6$ with the x-axis, and the curve $v = 0$ consists of a single branch with asymptotic directions 0 and $\pi/3$. Clearly $u = 0$ and $v = 0$ must intersect somewhere in the angular interval $(0, \pi/3)$. Fig. 6.2 shows all the

Fig. 6.1

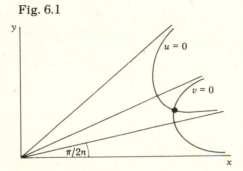

curves $u = 0$ and $v = 0$. There are exactly *three* points of inter-section, one of which lies on the negative imaginary axis. (In this simple case formulae can of course be obtained for the roots.)

The general argument we have outlined shows that a root exists but it does not give us a formula for it, and an existence proof may not help us actually to *compute* the object whose existence is proved (in the present problem a computer would always do it by numerical methods). For a modern version of Gauss's proof see Courant & Robbins (1941).

This result leads us on to the second question: having established their existence, can we not after all derive *formulae* for the roots of equations of degree greater than 4, which would allow us to calculate these roots by algebraic processes (i.e. by using the operations of addition, subtraction, multiplication, division and root extraction, as in Cardan's formula)? Or we might instead ask: *why* had no one been able to find such formulae? (After all, for equations of degrees 3 and 4 the process is fairly simple.) The answer to this question was found about 1820–30 and is interesting and important. Note that there are of course many special cases of higher-degree equations where formulae can be given for the roots (thus $x^n - a = 0$ has the n roots $x = a^{1/n}$; $x^6 - 2x^3 + 4 = 0$ is a quadratic equation in x^3, and we find $x^3 = 1 \pm \sqrt{(-3)}$, $x = [1 \pm \sqrt{(-3)}]^{1/3}$). But by about 1800

Fig. 6.2

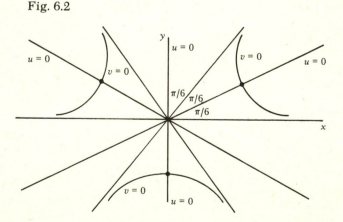

leading mathematicians such as Lagrange and Gauss suspected that it might be *in principle* impossible to solve the *general* equation of degree greater than 4 algebraically, i.e. that in the general case there is *no formula* which gives the roots algebraically in terms of the coefficients in the equation. But can one prove this, once and for all? Ruffini attempted such a proof (in 1799) but it was not conclusive. The matter was finally cleared up by two remarkable young geniuses, Abel and Galois.

N. H. Abel (1802–29) was the son of a Norwegian pastor who studied in Christiania (now Oslo) and Copenhagen and won a scholarship which took him to Paris. There he met the famous mathematicians of the age, but he was very shy and they ignored him. Eventually Abel returned to Norway, where he struggled to earn money to keep his family by giving lessons to young students. Just as his brilliant mathematical work began to attract attention he died of tuberculosis. Abel thought at one time that he had solved the equation of the fifth degree algebraically, but he corrected this claim in a paper which proved (in 1824) the *impossibility* of solving the general equation of the fifth degree by radicals. This celebrated theorem marked a great advance, but left open the question: just which equations are, and which are not, solvable by radicals; what exactly are the criteria for solubility? This and many related questions were answered by E. Galois (1811–32), the most romantic figure in the history of mathematics. He, also, made no impression on his mathematical contemporaries. Cauchy and Fourier lost the papers Galois submitted to them, and Poisson found them incomprehensible. He was also in trouble because of his political activities, and he died from wounds received in a duel. On the eve of the duel he jotted down his latest discoveries in a letter to a friend: 'you will publicly ask Jacobi or Gauss to give their opinion, not on the truth, but on the importance of the theorems. After this there will be, I hope, some people who will find it to their advantage to decipher all this mess.' (In fact Galois's last manuscripts were eventually published in 1846, but his work was not fully appreciated until 1870 or so.) The 'mess' was essentially the *theory of groups*; although earlier mathematicians had been aware, in a rudimentary way, of the concept of a group, it was

Galois who recognised its fundamental importance. In 1770 Lagrange, studying the solubility of algebraic equations by radicals, had found an important clue: he noted that the special tricks which work for equations of degree less than or equal to 4 all depend on finding functions of the roots of the equation which are unchanged under *permutations* of the roots, and he observed that this invariance property did not hold for $n = 5$. Galois associated with any polynomial a group of permutations of its zeros, now called the *Galois group*, and he showed that the problem of solubility by radicals can be reduced to a theorem on the structure of this group. Galois theory has many interesting applications in mathematics. In particular, it can be used to demonstrate the impossibility of trisecting an angle, or of 'duplicating a cube', by ruler-and-compass constructions, and thus it finally settles two of the 'three famous mathematical problems of antiquity' (see Chapter 8). For a modern account of Galois theory, with historical sections, see Stewart (1973).

We should not be surprised that Galois's work was not recognised in his own lifetime. Not only was he an impatient genius, unwilling to express himself in ways intelligible to more conventional mortals, but the emphasis on algebraic *structure* in his work only became part of the common way of mathematical thinking much later in the nineteenth century. All those who find the 'new mathematics' unfamiliar and strange will sympathise!

With the work of Gauss, Abel and Galois we know that an algebraic equation of degree n has n roots (in the field of complex numbers), and that in general no algebraic formula exists for the roots of an equation of degree greater than 4. Now let us recall the Greek discovery that the number $\sqrt{2}$ is irrational, i.e. that it cannot be written as the ratio of two integers. The number is nevertheless in a sense very closely related to the integers since it is the (positive) root of a simple algebraic equation (namely $x^2 - 2 = 0$) which has been formed from a polynomial with integer coefficients. The question arises: does *every* irrational number have this property, i.e. can irrational numbers always be obtained as the root of *some* equation of the form of equation (1)

$$x^n + a_1 x^{n-1} + a_2 x^{n-2} + \cdots + a_{n-1} x + a_n = 0,$$

where the as are integers? The roots of such an equation (real or complex) are *algebraic numbers,* so we ask: are there irrational numbers which are not algebraic? If such numbers exist, they must belong to a class with special and unusual properties.

In what ways, other than as solutions of algebraic equations, can irrational numbers arise? They often occur as sums of convergent infinite series; for example the famous number e ('Euler's number', the base of natural logarithms) is defined by

$$e = 1 + \frac{1}{1!} + \frac{1}{2!} + \frac{1}{3!} + \cdots .$$

Again, the number π, representing the ratio of the circumference of a circle to its diameter, can be defined by

$$\frac{\pi}{4} = \frac{1}{1} - \frac{1}{3} + \frac{1}{5} - \frac{1}{7} + \cdots ,$$

or by a definite integral:

$$\tfrac{1}{2}\pi = \int_0^\infty \frac{dx}{1+x^2},$$

or by an (infinite) 'continued fraction':

$$\frac{4}{\pi} = 1 + 1 \over 2 + 9 \over 2 + 25 \over 2 + 49 \over 2 + \cdots$$

(many other representations are possible). Note that in all these definitions an infinite process of some kind is involved.

It is quite easy to show that e is irrational (Euler, in 1737). Since

$$\frac{1}{2!} + \frac{1}{3!} + \frac{1}{4!} + \cdots < \frac{1}{2} + \frac{1}{2^2} + \frac{1}{2^3} + \cdots = 1,$$

we see from the series defining e that $2 < e < 3$. Now suppose that e is rational and hence equal to p/q, where p and q are integers and q must be greater than or equal to 2 (since e is not an integer). Then we must have

$$\frac{p}{q} = 1 + \frac{1}{1!} + \frac{1}{2!} + \cdots + \frac{1}{(q-1)!} + \frac{1}{q!} + \frac{1}{(q+1)!} + \cdots ,$$

so, multiplying by $q!$,

$$p(q-1)! = \left(q! + q! + \frac{q!}{2!} + \cdots + q + 1 \right) + \frac{1}{q+1}$$

$$+ \frac{1}{(q+1)(q+2)} + \cdots \tag{4}$$

The sum in brackets on the right-hand side is an integer, and

$$\frac{1}{q+1} + \frac{1}{(q+1)(q+2)} + \cdots < \frac{1}{3} + \frac{1}{3^2} + \frac{1}{3^3} + \cdots \text{ (since } q \geqslant 2),$$

and the sum of the geometric series on the right is $\frac{1}{3}/(1-\frac{1}{3}) = \frac{1}{2}$. Thus, in (4), the left-hand side is an integer but the right-hand side is not an integer, and we have a contradiction.

The number π is also irrational, but this is not so easy to prove. The proof was given by Lambert (in 1761) who showed that, if x is rational, then $\tan x$ *must* be irrational. Since $\tan \frac{1}{4}\pi = 1$, which is rational, $\frac{1}{4}\pi$ cannot be rational. It is often extremely difficult to decide whether any given number is rational or irrational. It has only very recently been shown that the number defined by the infinite series

$$\frac{1}{1^3} + \frac{1}{2^3} + \frac{1}{3^3} + \frac{1}{4^3} + \cdots$$

is irrational, and at the time of writing no one knows whether

$$\frac{1}{1^5} + \frac{1}{2^5} + \frac{1}{3^5} + \frac{1}{4^5} + \cdots$$

is rational or irrational.

Obviously all 'surds' formed by root extraction from whole number are algebraic numbers, for example $\sqrt[3]{(1+\sqrt{2})}$ is a root of the equation $x^6 - 2x^3 - 1 = 0$, and so on. But not all algebraic numbers are surds: the roots of the general polynomial equation (with integer coefficients) of degree greater than 4 are algebraic numbers, although, as we have seen, they are not in general expressible in terms of radicals. A number x is called an *algebraic number of degree n* if it satisfies an equation of the form (1), but no equation of lower degree: thus $\sqrt[3]{2}$ is an algebraic number of degree 3.

The existence of irrational numbers which are not algebraic was first demonstrated by Liouville (in 1844). Such numbers are called *transcendental*, because they 'transcend the power of algebraic methods' (Euler). Liouville actually *constructed* numbers which can be shown to be non-algebraic, for example

$$z = \frac{1}{10^{1!}} + \frac{1}{10^{2!}} + \frac{1}{10^{3!}} + \cdots .$$

This is a number defined by a series with extremely rapidly decreasing terms. Why is such a number not algebraic? The reason is, essentially, that an algebraic number can not be approximated by a rational number 'too closely'; specifically, if p/q is any rational approximation to an algebraic number x of degree n, with q a large positive integer, then it is not difficult to show (see Davenport 1970, p. 164) that the difference between x and p/q must satisfy the inequality

$$|x - p/q| > K/q^n,$$

where K is a constant for given x. Thus, for a given denominator q, the approximation p/q can never be closer than is permitted by this inequality. This result was proved by Liouville. Now, if we break off the infinite series which defines z after a sufficiently large number of terms, we have a rational approximation for z which is 'too close', i.e. the above inequality is violated. So transcendental numbers exist!† But, although this argument allows us to construct examples of such numbers, it is much more difficult to show that any particular *given* real number is transcendental. Liouville could not answer the question for e or π; the proof that e is transcendental was given by Hermite in 1873, and the proof for π was given by Lindemann in 1882. This last result, incidentally, finally and conclusively

---

† In Chapter 11 we mention another proof of the existence of transcendental numbers. It is more abstract but is perhaps easier to follow.

proved the impossibility of 'squaring the circle' with ruler and compass (since for this to be possible π would have to be an algebraic number).

It has also been shown that Liouville's inequality on the approximation of algebraic numbers by rationals can be strengthened considerably. This is a problem on which much work was done in the first half of this century, culminating in K. F. Roth's proof (in 1955) that

$$|x - p/q| > 1/q^{\nu}$$

for all but a finite number of approximations p/q, where ν can be any number greater than 2. This is the 'best possible' result, since the inequality does not hold for $\nu = 2$ (see Davenport 1970, p. 164).

The question of establishing the nature of numbers like $2^{\sqrt{2}}$ was posed in Hilbert's famous collection of outstanding unsolved problems in mathematics (in 1900). It was answered by Gelfond and Schneider in 1934 when they demonstrated the transcendence of α^{β}, where α is any algebraic number different from 0 and 1 and β is any irrational algebraic number. But what (for example) about α^{β} when α and β are both transcendental? The number $e^{\pi} = (-1)^{-i}$ is known to be transcendental (Gelfond), but at the time of writing the nature of the numbers e^{e} and π^{π} has not been definitely established. There are many other deep problems, too technical to be described here, in 'transcendental number theory' which is a highly active field of present-day research (see Baker 1975). The study of different types of numbers, with its long history going back to the Greeks, has not lost its fascination nor the ability to throw up very difficult mathematical problems.

7

Complex numbers, quaternions and vectors

Let us finally, in our discussion of the development of the number system, look again at the question of generalising the concept of number beyond the real numbers. We recall that, to solve an algebraic equation with real coefficients such as $x^2 + 1 = 0$, it was necessary to introduce the complex numbers $x + iy$, but the set of complex numbers is 'algebraically closed'. Every algebraic equation with real or complex coefficients has solutions in the field of complex numbers. This implies, in particular, that roots and powers of complex numbers always exist within the field of complex numbers. So, as far as the solution of algebraic equations of all kinds is concerned, our search is finished – no further generalisation of number is needed. But in another, more geometrical, sense the search is not finished. We discussed in Chapter 3 the interpretation of complex numbers as displacements, or vectors, in a space of two dimensions (the complex plane), and we noted that multiplication by a complex number of unit magnitude could be regarded as a rotation in this space. In classical applied mathematics we usually deal with displacements and vectors in 'physical' space which is three-dimensional. It is therefore a natural question, and one likely to be of practical importance, to ask whether one can define 'complex numbers' in spaces of dimension greater than two which, one would hope, could represent displacements and rotations in such spaces. In particular, we want to know whether we can invent a 'hypercomplex number' which can represent a vector in three dimensions and which is such that multiplication by it is equivalent to a rotation in three-dimensional space. Such an entity would be extremely useful for the formulation of the laws of physics, and in particular for the description of the motion of rigid bodies in space. Also, quite apart from its physical applications, the search for generalisations of the notion of complex number to spaces of many dimensions is a question of intrinsic mathematical interest.†

The answer to this question is less simple than one might at first sight suppose. Once an understanding of the two-dimensional geometrical representation of complex numbers had been achieved in the early nineteenth century, the search for a generalisation to three and more dimensions was taken up by many mathematicians. A key step forward was taken by William Rowan Hamilton (1805–65), who did brilliant work both as a mathematician and as a physicist. Hamilton was an infant prodigy (at the age of 13 he was said to have been familiar with as many languages as he had lived years). At Trinity College Dublin he was appointed professor of astronomy while still an undergraduate. His early work included a highly original general reformulation of the laws of optics and mechanics. The 'Hamiltonian function' plays a fundamental role in modern quantum theory. Not all of Hamilton's ideas have stood the test of time, however. One of his favourite themes was that 'space and time are indissolubly connected'; much later on this notion was indeed incorporated in the theory of relativity, but Hamilton argued that, since geometry is the 'science of space', algebra must be the 'science of pure time'; such an idea does not have any meaning for us.

In 1833 Hamilton wrote a paper in which complex numbers were regarded as *ordered pairs* (a, b) of real numbers, with the rules for addition and multiplication given by

$$(a, b) + (c, d) = (a + c, b + d),$$
$$(a, b)(c, d) = (ac - bd, ad + bc);$$

and he recognised that these algebraic rules implied the interpretation of addition as (parallelogram) addition of vectors, and

† We give no formal definitions here of notions such as 'space', 'rotation' and 'dimension'; it will be sufficient for the present to regard an n-dimensional space as a generalisation of 'physical' (Euclidean) three-dimensional space, with the coordinates of a 'point' given by n real numbers $x_1, x_2, \ldots, x_n$. The notion of 'dimension' in fact contains subtleties which are discussed in Chapters 10 and 11.

of multiplication by $(\cos\theta, \sin\theta)$ as a rotation. In this formulation a complex number was for the first time explicitly recognised as equivalent to an ordered pair of real numbers, and such a pair was seen to be equivalent to a two-dimensional vector.

This point of view led Hamilton to consider, as the natural generalisation to three dimensions, *ordered triples* (a, b, c) of real numbers. Or, equivalently, a complex number written as $a + bi$ would be replaced by something of the form $a + bi + cj$, containing now *two* non-real 'units' i and j replacing the 'imaginary unit' i. Such a generalised complex number will certainly represent a vector in three-dimensional space, with rectangular components a, b, c, and there is no difficulty over defining *addition* for these entities: the rule is

$$(a + bi + cj) + (a' + b'i + c'j)$$
$$= a + a' + (b + b')i + (c + c')j,$$

and this is just the general parallelogram law of addition for forces, which had been known from ancient times. However, Hamilton was held up for ten years in following up this idea because of difficulties with the definition of *multiplication* for these generalised complex numbers and its relation to rotations in three-dimensional space. It is said that in 1843, when he was walking with his wife along the Royal Canal in Dublin, Hamilton suddenly saw what had to be done: to represent rotations in three dimensions he must use *quadruples* of real numbers, not triples, and he must also abandon the *commutative law of multiplication*!

It is in fact clear on geometrical grounds that four real numbers, not three, are needed. In two dimensions we need two numbers, because multiplication by a complex number must in general specify both an angle of rotation and the ratio by which the length of the vector is changed. In three dimensions, if multiplication by our 'hypercomplex number' is to transform the vector A into some other vector B (Fig. 7.1), we must specify first the *direction* in space of the axis OO' about which the vector A is to be rotated: specification of a direction in space needs two numbers (for example the latitude and longitude angles). Then we need another two numbers to specify the angle of rotation about the axis and the 'stretch' of the vector, making four in all. Thus our numbers must be of the form $a + bj + ck + d\ell$, where a, b, c, d are four real numbers and j, k, ℓ are new 'unit elements' analogous to i for ordinary complex numbers. Hamilton called these new entities *quaternions*.

What rules must be adopted for calculating with j, k, ℓ (replacing the rule $i^2 = -1$)? At first Hamilton assumed that the 'ordinary' rules of traditional algebra must apply (we would say nowadays that the new numbers must form a field); we must remember that it had not really occurred to anyone before his time that there *could* be any other rules in algebra. In fact the notion that, in the search for generalisation, the 'familiar' rules of calculation must continue to apply had been elevated into a mathematical principle, the *principle of continuity* (or *permanence*): its formulation was vague, but it was a serious hindrance to progress.

Hamilton found that application of the familiar rules of addition and multiplication to his quaternions invariably led to a contradiction somewhere. It was his inspiration to see that these rules, derived from experience in calculating with 'ordinary' numbers, were not the only ones possible, and that in calculating with quaternions 'something' must be given up. Of course the step from real to complex numbers had already been a

Fig. 7.1

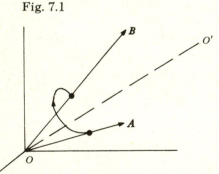

move towards abstraction, but now a much bolder innovation was needed. Hamilton realised that, if his quaternions were to follow rules 'as similar as possible' to the familiar ones, it was the *commutative law* of multiplication which must be abandoned. This becomes clear when we try to define *division* by a quaternion.† To do this we require that, given any non-zero quaternion w, the equation $ww_1 = 1$ (where '1' is the *unit quaternion* $1 + 0j + 0k + 0\ell$) should have a unique quaternion solution w_1 (then we can write $w_1 = w^{-1}$ and define division by w as multiplication by w_1). How does such an approach work for ordinary complex numbers? Since $a^2 + b^2 = (a + ib)(a - ib)$, every non-zero complex number $a + ib$ has a unique complex number as its inverse:

$$(a + ib)^{-1} = \frac{1}{a + ib} = \frac{a - ib}{a^2 + b^2}$$

$$= \left(\frac{a}{a^2 + b^2}\right) - i\left(\frac{b}{a^2 + b^2}\right).$$

($a - ib$ is the complex number *conjugate* to $a + ib$, and the positive real number $a^2 + b^2$ is the *norm* of the complex number $a + ib$.)

To do the same thing for the quaternion $w = a + bj + ck + d\ell$ we need similarly to factorise the norm $N(w)$, now defined as the positive real number $a^2 + b^2 + c^2 + d^2$, into a product of two linear quaternion factors. To achieve this let us try

$$a^2 + b^2 + c^2 + d^2 = (a + bj + ck + d\ell)(a - bj - ck - d\ell),$$

where we regard $a - bj - ck - d\ell$ as the conjugate of $a + bj + ck + d\ell$. If we multiply out the right-hand side, using the 'ordinary' rules, we obtain terms like $-b^2j^2$ (showing that $j^2 = -1$), but also 'cross terms' like $-2bcjk$; thus the factorisation fails (since clearly j, k are both non-zero). But, if the commutative law of multiplication does not hold for j, k, ℓ, so that $jk \neq kj$, etc., then the *order* in which the terms appear in the multiplication matters, and a typical cross term must be written $-bc(jk + kj)$; it

---

† We are not following here precisely Hamilton's own train of reasoning. He explains this himself in the preface to his *Lectures on Quaternions* (see Hamilton 1967).

will now be zero if $jk = -kj$. Clearly we achieve the desired factorisation by taking

$$j^2 = k^2 = \ell^2 = -1$$

and

$$jk = -kj, \qquad k\ell = -\ell k, \qquad \ell j = -j\ell.$$

To ensure that the product of any two quaternions will again be a quaternion we require in addition that a product such as jk should be a linear combination of j, k and ℓ. In fact Hamilton postulated that

$$jk = -kj = \ell, \qquad k\ell = -\ell k = j, \qquad \ell j = -j\ell = k.$$

With these rules we now have

$$(a + bj + ck + d\ell)^{-1} = \frac{a}{r} - \frac{b}{r}j - \frac{c}{r}k - \frac{d}{r}\ell,$$

where

$$r = a^2 + b^2 + c^2 + d^2,$$

and the quaternions satisfy all the axioms for a field *except* that multiplication is in general not commutative. (Such a set is nowadays called a *skew field* or *division algebra*.) So, for a product of any two quaternions, $w_1 w_2 \neq w_2 w_1$ in general. But note that our argument implies that the norm of the product, $N(w_1 w_2)$, is equal to the product of the norms $N(w_1)N(w_2)$ (you can verify this by direct calculation of the norms). We say that the division algebra is *normed*.

To finish the story about Hamilton: it seems that he was so excited by his discovery that he took out a knife and cut the fundamental formulae $j^2 = k^2 = \ell^2 = jk\ell = -1$ into the stone of the bridge. Hamilton's invention of quaternions (the first non-commutative algebra) was a revolutionary step, as it shattered preconceived notions as to how numbers 'must' behave. It was at about the same time that people such as George Peacock in Cambridge first tried to formulate systematically the fundamental laws of arithmetic and algebra in axiomatic form. The early attempts at identifying these laws were not very satisfactory, since intuitive notions and preconceived habits of thought about the way mathematical entities 'must' behave were still too strong (we have already mentioned the 'principle of continuity', a blind alley). After the work of Hamilton, and others at about the same time, it began to be realised that the *rules* of algebra could be formulated independently of the *objects* to which they were applied; this discovery set algebra free to invent and study new structures, and a very rapid development set in around the middle of the nineteenth century. One of the most original contributions was Grassmann's *Lineale Ausdehnungslehre* (Theory of Linear Extension, published in 1844), a very general vector algebra, highly abstract and written in rather obscure form and therefore slow to be recognised. Another important development was the axiomatic formulation of the laws of formal logic by George Boole in his famous books, *The Mathematical Analysis of Logic* (1847) and *Investigation of the Laws of Thought* (1854). Boolean algebra is a general algebra of sets which is of wide application to modern problems, in particular to probability problems arising in business and insurance and to the switching system of a large digital computer.

Returning to quaternions, let us now examine how they in fact describe rotations of vectors in three-dimensional space. Note first that the need for a non-commutative multiplication law again becomes evident when we think about rotations, because the combination of two rotations about different axes, carried out in succession, is a non-commuting process: the result depends on the order in which the rotations are carried out. To convince yourself of this, let Ox, Oy, Oz be three mutually perpendicular axes. Take a pencil oriented along Oz, rotate it through a right angle about Ox so that it lies along Oy, and then rotate it through a right angle about Oz so that it finishes along Ox. If you rotate *first* about Oz and *then* about Ox, the pencil will finish up along Oy!

We represent a three-dimensional vector v, with components b, c, d along Ox, Oy, Oz, by the quaternion $b\mathrm{j} + c\mathrm{k} + d\ell$ (this is sometimes called a 'pure quaternion'; it has no 'real part'). We want to rotate this vector through a given angle θ about a given axis whose direction can be specified by its *direction cosines* α, β, γ (these are the cosines of the angles between the axis and

Ox, Oy, Oz; they satisfy $\alpha^2 + \beta^2 + \gamma^2 = 1$). The rule is that the new vector $v' = b'\text{j} + c'\text{k} + d'\ell$ is obtained by forming the quaternion product

$$v' = \{\cos \tfrac{1}{2}\theta + \sin \tfrac{1}{2}\theta(\alpha\text{j} + \beta\text{k} + \gamma\ell)\}$$
$$\times v\{\cos \tfrac{1}{2}\theta - \sin \tfrac{1}{2}\theta(\alpha\text{j} + \beta\text{k} + \gamma\ell)\}.$$

(To make the rule quite unambiguous we must also specify the *sense* of the rotation about the axis. θ is *positive* if the rotation is seen as *anticlockwise* by someone observing it from the 'positive end' of the axis of rotation; see Fig. 7.2.)

We will not give a general derivation of this formula, but let us check it for the special composition of two rotations mentioned above. Suppose the original vector lying along Oz was of unit length; then $v = \ell$. Rotate v about the x-axis through $+\tfrac{1}{2}\pi$: then $\theta = \tfrac{1}{2}\pi$, $\alpha = 1$, $\beta = \gamma = 0$, and the rotated vector is

$$(\cos \tfrac{1}{4}\pi + \text{j} \sin \tfrac{1}{4}\pi)\ell(\cos \tfrac{1}{4}\pi - \text{j} \sin \tfrac{1}{4}\pi) = \tfrac{1}{2}(1+\text{j})\ell(1-\text{j}),$$

since $\cos \tfrac{1}{4}\pi = \sin \tfrac{1}{4}\pi = 1/\sqrt{2}$. Using the quaternion multiplication laws we find that this becomes $\tfrac{1}{2}(\ell - \text{k})(1 - \text{j}) = \tfrac{1}{2}(\ell - \text{k} - \text{k} - \ell) = -\text{k}$. Now rotate about the z-axis through $\tfrac{1}{2}\pi$: we obtain

$$(\cos \tfrac{1}{4}\pi + \ell \sin \tfrac{1}{4}\pi)(-\text{k})(\cos \tfrac{1}{4}\pi - \ell \sin \tfrac{1}{4}\pi) = -\tfrac{1}{2}(1+\ell)\text{k}(1-\ell)$$
$$= -\tfrac{1}{2}(\text{k}-\text{j})(1-\ell) = -\tfrac{1}{2}(\text{k}-\text{j}-\text{j}-\text{k}) = \text{j};$$

thus we finish with a vector along Ox. Now reverse the order of rotation, and we obtain first

$$(\cos \tfrac{1}{4}\pi + \ell \sin \tfrac{1}{4}\pi)\ell(\cos \tfrac{1}{4}\pi - \ell \sin \tfrac{1}{4}\pi) = \ell,$$

and then

$$(\cos \tfrac{1}{4}\pi + \text{j} \sin \tfrac{1}{4}\pi)\ell(\cos \tfrac{1}{4}\pi - \text{j} \sin \tfrac{1}{4}\pi) = -\text{k}.$$

Fig. 7.2

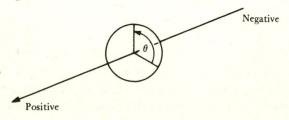

Negative

Positive

We see that the different end positions in the two cases are a consequence of the non-commutative multiplication law for quaternions. We can deal similarly with any sequence of rotations in three-dimensional space.

Our quaternion rule for rotating vectors in three dimensions looks quite different from the simple multiplication of complex numbers which does the trick in two dimensions. The reasons for this difference are of considerable mathematical interest. A full exploration would lead us into advanced theory, but we may note the following. Our rotations are constructed with the use of the quaternion $\cos \frac{1}{2}\theta + \sin \frac{1}{2}\theta(\alpha j + \beta k + \gamma \ell)$ and its conjugate. This quaternion, like the complex number $\cos \theta + i \sin \theta$, has *unit norm* (since $\cos^2 \frac{1}{2}\theta + \sin^2 \frac{1}{2}\theta(\alpha^2 + \beta^2 + \gamma^2) = 1$). It corresponds to a point on the surface of the *unit sphere* S^3 in a space of *four* dimensions. A continuous rotation in three dimensions can thus be recorded by plotting the (continuous) path on S^3 traced out by the corresponding quaternion. But, because of the appearance of the half-angles $\frac{1}{2}\theta$ in the rotation quaternion, a given rotation through θ corresponds, not to one, but to *two* points on S^3 lying at opposite ends of a diameter (the angles θ, $\theta + 2\pi$ give the same rotation in three dimensions, but the quaternion changes sign when we replace θ by $\theta + 2\pi$). Thus we need *two* rotations through 2π in three dimensions in order to return to our starting point on S^3. This curious behaviour is connected with the fact that the path of the rotation on S^3 (a great circle) can be *continuously deformed* into a point on the sphere; this is something that has no analogue in the two-dimensional case where our paths are confined to the unit circle S^1, and a great circle on S^1 always coincides with S^1 itself and cannot be deformed. The study of such continuous deformations belongs to the subject of *topology*, a vast and highly active branch of modern mathematics which is becoming increasingly important in many applications to science.

Hamilton had an exaggerated enthusiasm for his quaternions and spent the rest of his life working on them. He thought that his discovery was as important as that of the infinitesimal calculus, and he regarded it as the key to all of geometry and mathematical physics. A few scientists in England (particularly

P. G. Tait) shared Hamilton's enthusiasm and did their best to propagate the use of quaternions, but on the whole the concept was ignored by physicists who went on using Cartesian coordinates in their theories. In fact quaternions were not what physicists really wanted; they were looking for a concept more directly associated with the three Cartesian coordinates representing a three-dimensional vector. For mathematicians nowadays quaternions represent just one example of an algebraic system, and there are very many other systems all with their own interesting properties. Quaternions, and other hypercomplex number systems, owe their continued mathematical interest chiefly to their connection with topology to which we have alluded above.

We mention briefly some developments in algebra in the second half of the nineteenth century which followed Hamilton's discovery of quaternions.

Pursuing the argument which led to quaternions we may ask whether we can generalise the concept of number further, by inventing new hypercomplex numbers in a greater number of dimensions. In 1845 Cayley (and independently Hamilton's friend J. T. Graves) showed that there is an *eight-dimensional* generalisation of number ('octonions', constructed from pairs of quaternions) which satisfies the rules for a field except that for these entities multiplication is in general *neither* commutative *nor* associative (so that $(ab)c \neq a(bc)$ in general). We have here again a 'normed division algebra' in which every non-zero 'number' has an inverse. These 'Cayley numbers' are little more than mathematical curiosities. But are there any other 'division algebras'? This question poses a very difficult problem which required the full resources of modern algebraic topology for its solution. It was finally answered (with *no*) in 1958, when J. F. Adams and others showed that there are no division algebras constructed from the real numbers other than those of dimension 1, 2, 4 or 8. This shows again that quaternions and octonions are of only restricted significance in algebra.

But suppose we ask about the description of *rotations* in general spaces. (Note that in the nineteenth century the idea that geometry might make sense in more than three dimensions was, once again, a strange new concept, at first treated with suspicion and disbelief, and slow to be generally accepted. The same was true, of course, of *non-Euclidean geometry*, in which Euclid's 'parallel postulate' was replaced by a different axiom about the number of lines that could be drawn through a given point parallel to a given line. Non-Euclidean geometry dates from about 1830 and was thus historically the earliest break away from 'traditional' assumptions about the proper concerns of mathematics.) In four dimensions we can again use quaternions to carry out rotations: the general quaternion $a + b\mathrm{j} + c\mathrm{k} + d\ell$ constructed from four real numbers can be regarded as a vector in four dimensions, and a generalisation of our earlier rule for rotating vectors in three dimensions describes rotations of such vectors in four-dimensional space. An extension of the quaternion calculus to the general case of n dimensions which can be used to describe rotations of n-dimensional vectors was given by W. K. Clifford in 1878. The *Clifford algebras*, again, play a significant role in modern differential topology. Rotations in four dimensions are important in the theory of relativity (see below, p. 85).

To define the 'algebra' of vectors in n dimensions one needs to know the 'multiplication table' (analogous to $\mathrm{jk} = -\mathrm{kj} = \ell$, etc., for quaternions) of all possible products $e_j e_k = \sum_{i=1}^{n} c_{ijk} e_i$ which can be formed from the n unit vectors e_i $(i = 1, 2, \ldots, n)$ in the space considered. If we require in addition that, always, $(e_i e_j)e_k = e_i(e_j e_k)$, we have a *linear associative algebra*. The general theory of such algebras was investigated in the 1860s by the American mathematician Benjamin Peirce and his son Charles. Benjamin Peirce worked out the multiplication tables for 162 different algebras, a far cry from the assumption in the early nineteenth century that there could be only *one* set of rules for 'algebra'!

Our topic is the concept of 'number'. With algebra's new-found freedom to invent its own rules it became largely a matter of definition and language to decide which entities should be called 'numbers' (clearly one would expect them to have some, at least, of the traditional properties of numbers). It is important to realise that the new freedom also led, from around 1850 onwards, to increasing interest in mathematical structures obeying less restrictive laws than those traditionally associated with numbers. It gradually became clear to mathematicians that they could obtain structures of very great mathematical interest by imposing *fewer* rules on the elements of their collection (or *set*) than the full array of arithmetical operations. Next to the general theory of sets, perhaps the most fundamental concept to emerge was that of the *group*, already mentioned in Chapter 6, in which one has only one binary operation which assigns to any pair of elements in the set another element in the same set.† In a modern mathematics course non-commutative multiplication is often first encountered in connection with the arrays of numbers called *matrices*. They arise when one is studying the linear transformations which change a vector into another, and it was Cayley again who first studied matrices systematically (1858) in the context of linear transformations in geometry. In the calculus of square matrices one has two laws of combination (addition and multiplication) and multiplication is associative, but matrices have a fundamentally new property not possessed by numbers (elements of a field): the relation $AB = 0$ (where '0' is the zero matrix) can hold for two matrices even when A and B are both non-zero.‡ Non-zero matrices whose product is zero do not possess inverses: for a matrix A of this type one cannot find a matrix A_1 such that AA_1 is equal to the unit matrix. They are called *singular* matrices.

Since matrices and quaternions are both associated with rotations, one might expect a connection between them. This exists, and in fact the algebra of Hamilton's quaternion units j, k, ℓ is precisely the same as that of the following special 2×2

---

† We emphasise again that we are not attempting here to give full definitions; these will be found in any textbook of algebra. It should also be realised that concepts such as *group, ring, field* etc. emerged gradually during the nineteenth century, and their general significance was only slowly recognised: general abstract definitions (with which modern treatments usually start) only came later.

‡ An example (for those familiar with matrix multiplication):

$$\begin{pmatrix} 1 & 2 \\ 2 & 4 \end{pmatrix}\begin{pmatrix} -2 & -4 \\ 1 & 2 \end{pmatrix} = \begin{pmatrix} 0 & 0 \\ 0 & 0 \end{pmatrix}.$$

The set of all 2×2 matrices is said to form a *ring*.

matrices built from complex numbers:

$$\sigma_1 = \begin{pmatrix} 0 & i \\ i & 0 \end{pmatrix}, \qquad \sigma_2 = \begin{pmatrix} 0 & 1 \\ -1 & 0 \end{pmatrix}, \qquad \sigma_3 = \begin{pmatrix} -i & 0 \\ 0 & i \end{pmatrix},$$

with $i^2 = -1$. Using the rules of matrix multiplication you will quickly verify that $\sigma_1^2 = \sigma_1 \times \sigma_1 = -I$

$$\left(\text{where } -I \text{ is the matrix } \begin{pmatrix} -1 & 0 \\ 0 & -1 \end{pmatrix} \right),$$

and that $\sigma_1\sigma_2 = -\sigma_2\sigma_1 = \sigma_3$, etc.; these are precisely the rules for j, k and ℓ. We say that the general non-zero quaternion $a + bj + ck + d\ell$ is *isomorphic* with the 2×2 complex matrix defined by

$$a1 + b\sigma_1 + c\sigma_2 + d\sigma_3 = \begin{pmatrix} a - id & c + ib \\ -c + ib & a + id \end{pmatrix}.$$

(A non-zero matrix of this particular form is non-singular and has an inverse, so there is no contradiction with what was said about quaternions.) These 2×2 complex matrices (or the corresponding quaternions) can be regarded as describing rotations of two-dimensional complex vectors called *spinors*, and the matrices $\sigma_1, \sigma_2, \sigma_3$ (multiplied by $-i$) are the *Pauli spin matrices*. These names derive from the use of spinors in modern quantum theory to describe the state of an electron, a particle with *spin* (intrinsic angular momentum). The appearance of the half-angles in the quaternions which describe the rotations of spinors, and the resulting peculiar double-valued property which we discussed earlier, here acquires physical significance; it is connected with the fact (a quantum-theoretical property) that the electron's spin is half-integral. The theory of the electron spin was put forward in the 1920s, and this application of his quaternions could not have been foreseen by Hamilton!

But, as we have already noted, Hamilton's quaternions did not provide the all-purpose vector calculus needed for classical physics. Hamilton himself and some later nineteenth-century physicists, in particular Clerk Maxwell (famous for his theory of the electromagnetic field), tried to overcome the difficulty by treating the 'real' and the 'pure quaternion' parts of a quaternion separately (see p. 75): these were then called the *scalar* and *vector* parts of the quaternion. For instance Hamilton had noted that, if $\alpha = b\mathrm{j} + c\mathrm{k} + d\ell$, $\alpha' = b'\mathrm{j} + c'\mathrm{k} + d'\ell$ are two pure quaternions, the 'scalar part' of the product $\alpha\alpha'$ is $-(bb' + cc' + dd')$ and the 'vector part' is

$$(cd' - dc')\mathrm{j} + (db' - bd')\mathrm{k} + (bc' - cb')\ell,$$

and he had discussed the geometrical implications of these formulae; he had thus already in the 1840s exhibited the structure of the scalar and vector products in modern vector algebra. In such ways the algebra of quaternions can be developed so as to exhibit many of the results of modern vector analysis, but it was eventually recognised that the quaternion formalism is unnecessarily cumbersome. The modern algebra of three-dimensional vectors which has become a standard mathematical language in physics and geometry was created in the 1880s by two people: J. W. Gibbs in the USA, a physicist, and Oliver Heaviside in England, an electrical engineer. They broke away completely from the concept of quaternions and treated the 'vector part' as an entirely separate entity. The 'vector algebra' which was then created for calculating with these objects is a branch of mathematics originally developed for use in science. Vectors have a more complex algebraic structure than do quaternions and other hypercomplex numbers. For example we define two kinds of product (useful in three dimensions): the 'vector product' which is non-commutative and analogous to the quaternion multiplication rule $\mathrm{jk} = -\mathrm{kj} = \ell$; and the 'inner product' (or 'scalar product') which is a number, not a vector. The concept of a product which does not belong to the original collection is a new idea, not contained in the structure of a field.

This work on vectors led to bitter controversy with the adherents of quaternions over the relative merits of the two algebras. In 1895 two colleagues of Gibbs at Yale University felt it necessary to organise an 'International Association for Promoting the Study of Quaternions and Allied Systems of Mathematics'. This published its own bulletin for a while (but, it seems, did not in fact spend much time in 'promoting' quaternions). Nowadays we no longer get heated about the matter, and we know that both vectors and quaternions have their recognised place in science and mathematics. In a way the vectors have won the argument, because vector algebra embodies an idea, that of *linearity* and *linear superposition*, which has turned out to be of very general significance in mathematics and many of its applications. Thus vectors are nowadays widely used throughout mathematical science. The basic laws of mechanics and electromagnetism are relations between three-dimensional vector quantities: forces, accelerations, electric and magnetic fields. In physics these vectors now form part of the more general mathematics of *tensors* (quantities with special linear transformation properties). The tensor calculus was originally developed for geometry by Riemann and later by Ricci and Levi-Civita (Italy). It is the formalism required for studying *symmetry* and *invariance* properties of geometric and physical magnitudes under various kinds of transformation. Tensors thus find applications in the study of crystal lattices, in elasticity (hence their name), and above all in Einstein's general theory of relativity (1915) which underlies our present-day efforts to understand gravitation and many of the large-scale properties of the universe. Modern analysis and geometry in n dimensions also make extensive use of the tensor concept.

The basic idea of linear superposition has given rise to the modern mathematical concept of a *linear vector space*, where 'space' and 'vector' are defined as abstract mathematical entities. The idea derives from the physicist's vector addition of forces to give a resultant force. The mathematician conceives a linear vector space L (over a field K) to be a set of abstract elements $a, b, c, \ldots$ (the 'vectors') for which addition is defined, together with elements $\alpha, \beta, \ldots$ of a field K (the 'scalars', for example the field $\mathbb{C}$ of complex numbers). Multiplication of vectors by scalars is defined and gives new vectors such as αa, βb and the linear combination $\alpha a + \beta b$: the key idea is then that $\alpha a + \beta b$ also belongs to the set L. This has become an important unifying idea in mathematics. Thus in analysis the set of all twice-differentiable functions $f(x)$ of the real variable x forms a linear vector space (a *function space*) over the real or complex numbers. This realisation leads to a theory of linear differential equations in which functions are treated as vectors. Function spaces have an infinite number of dimensions, and a full investigation of their mathematical properties is difficult, but function-vectors in such spaces have many important applications: for example they are the fundamental ingredients in the mathematical formulation of quantum theory. In such ways, as always in the past, mathematical and scientific ideas continued to influence each other, backwards and forwards.

It is thus not surprising that the modern mathematics student hears a lot about vectors in various parts of his course. He should perhaps bear in mind that the emphasis and the motivation for the study of these objects are not always the same. In algebra it is the general properties which are of chief interest, and the theory is formulated with the least possible restrictions. Therefore the basic axioms do not include any products of two vectors: these are not part of the basic idea of the linear combination of objects. When we are doing physics and analytic geometry in three dimensions we want our vectors to represent physical magnitudes and displacements, and we would like them to have both a *length* (a positive real number, not a vector) and a *direction* in space (so that we can ask for the *angle* between two vectors). Therefore we make additional postulates which allow us to define these attributes. We then have a *Euclidean* vector space in which the length of a vector is defined in terms of the scalar product of two vectors. In three-dimensional Euclidean space the square of the length of the displacement vector which joins the origin to the point P with coordinates (x, y, z) relative to a set of mutually perpendicular axes is $x^2 + y^2 + z^2$, in accordance with the ancient theorem of Pythagoras. This quantity (equivalent to what we earlier called a 'norm') remains invariant when we rotate the coordinate axes about the origin.† Although this definition of the length (or magnitude) of a vector is the most familiar, other definitions are possible and may be of interest in applications. An important

---

† This means that $x^2 + y^2 + z^2 = x'^2 + y'^2 + z'^2$, where (x', y', z') are the coordinates of the point P referred to a rotated set of perpendicular axes. Instead of considering a given *vector* and rotating the axes, we can consider the rotation of a vector in a given *coordinate system*, as described for example by the quaternion multiplication rule on p. 76. The reader may verify that, under this rule, $b^2 + c^2 + d^2 = b'^2 + c'^2 + d'^2$, so that the length of the vector is invariant.

generalisation occurs in the special theory of relativity where we are interested in *events*, characterised by their coordinates (x, y, z, t) in both space and time, and thus represented by points in a *four-dimensional* mathematical space (the 'space–time world'). Transformations of axes in this space (called *Lorentz transformations*) connect the space and time coordinates of an event as seen by observers in uniform motion relative to each other; and the fundamental fact that the velocity of light c is the same for all such observers is expressed by the mathematical requirement that the quadratic form $x^2 + y^2 + z^2 - c^2 t^2$ must be invariant under Lorentz transformations. This 'space–time norm' can be either positive or negative for real x, y, z, t (we say that it is 'indefinite'), and the Lorentz transformations of special relativity leave such an *indefinite* quadratic form invariant. There is a close connection between the 'four-vectors' of special relativity and the two-dimensional spinors mentioned on p. 81: it was Dirac's great discovery (in 1928) that the relativistic quantum theory of the electron automatically describes a particle possessing half-integral spin.

8

Greek ideas about infinity

The concept of infinity lies at the core of mathematics. We have already met it many times in our discussion of the idea of number. Even in the very simplest case, the collection of natural numbers 1, 2, 3, . . . , the series does not end and we have an infinitely large collection.

The Babylonians must have encountered infinite processes, but they avoided coming to grips with them. For example, when working out their sexagesimal fractions, they made a table of reciprocals which gives the reciprocal of 10 as the finite fraction 6/60, the reciprocal of 12 as the finite fraction 5/60, but the reciprocal of 11, which is an infinite series, is simply omitted from the table.

Not only the idea of infinitely large collections, but also that of infinitely small subdivisions, created difficulty from early days. The infinitely small arose in two questions of great concern to the Greeks: the problem of finding the areas of figures bounded by curved lines, and the problem of understanding motion, in particular the motions of the heavenly bodies studied by the astronomers. Why do such questions involve the infinitely small? In the case of areas there is no difficulty so long as we are dealing with plane figures bounded by straight lines. Thus, for a square, since *area* is itself defined as *rectilinear* area, i.e. the number of squares of unit size contained in the figure, we need only subdivide the area into unit squares and count their number, provided always that we have a unit of length in terms of which we can express the length of a side (Fig. 8.1). And this notion is easily extended to rectangles, triangles and polygons. But what happens when the boundary is 'continuously turning'? Take the area of a circle; we may draw an inscribed polygon to give us an approximate estimate of the area (Fig. 8.2), and we may then increase the number of sides of the polygon to improve the approximation. But does this process continue indefinitely? Or can we stop at some stage and say that our polygon now coincides with the circle? If so, how far must we go? If not, we must presumably continue until we have 'infinitely many, infinitely small' sides; what does this mean? We would now call this puzzle a problem in *integral calculus*, in contrast to the differential calculus which deals with problems of rates of change and tangents to curves. The integral calculus is in fact the older branch of the 'infinitesimal calculus', although the differential calculus is usually taught first nowadays. Again, when we come to the study of motion, a body is continuously changing its position with time; how does this happen? Can we think of time as infinitely subdivisible, or is there a smallest unit of time, so that time jumps 'discontinuously' from one instant to the next? The problem of infinite subdivisibility troubled the Greeks very much, and there were two main philosophic schools of thought: the school which believed in infinite subdivisibility and denied the existence of any smallest units, and the school of *atomism* which held that everything in our experience (in particular matter, time and space) was ultimately made up of smallest basic units. The problem of ultimate constituents is the most fundamental one in science. It is far from finally decided at the present time, although in the case of the constitution of matter

Fig. 8.1

Fig. 8.2

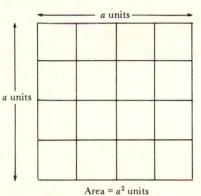

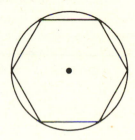

Area = a^2 units

we nowadays hold with atomism, while in our notions of (physical) space and time we believe in continuity and infinite subdivisibility. Since both atomism and continuity are *models* of empirical reality there is no final certainty possible on this question. While the basic philosophic problems must remain unresolved, we can at least claim that we have much more powerful ways these days of formulating these old questions mathematically.

The Pythagoreans reduced all problems to whole numbers and thus believed in numerical atomism! The *Eleatics* were a rival school of philosophers who believed in the so-called 'permanence of being'; for them all change was only apparent. Zeno (about 450 B.C.), their best-known member, propounded the famous 'paradoxes on motion'. These propositions were designed to show that motion was impossible under both types of assumption. Thus the 'Achilles paradox', which assumes infinite subdivisibility of space and time intervals, claims to show that the fast Achilles can never overtake the slow tortoise in a race in which the tortoise has a start and both start running simultaneously (Fig. 8.3): if A runs twice as fast as T, then, when A has reached A_2 (level with T_1), T will be at T_2, where $T_1T_2 = \frac{1}{2}A_1A_2$; when A is at A_3 (level with T_2), T will be at T_3 where $T_2T_3 = \frac{1}{2}A_2A_3$; and so on indefinitely. Thus T is always ahead of A. Zeno did not really believe that in practice Achilles would never overtake: his argument challenged the common belief that the sum of an infinite number of quantities can be made as large as we like, and showed in a vivid way that a proper logical theory of infinite series was needed to resolve the problem. The paradox of the 'Stade' (stadium) uses the concept of relative motion to demonstrate that we also get into difficulties with the idea of a smallest unit of time (see Boyer 1968, p. 83). Suppose B moves to the right relative to A with a speed such that

Fig. 8.3

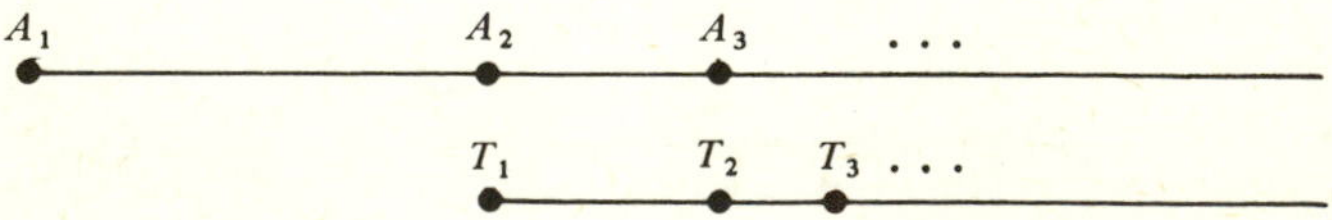

one unit of B (such as b_4) passes one unit of A (such as a_4) in the time unit τ (Fig. 8.4). Similarly C moves to the left with the same speed. Starting in configuration (a) we reach configuration (b) in time τ. But now c_1, originally opposite b_3, is opposite b_1, having passed *two* B units; hence τ cannot be the smallest possible time interval, since we can take as a new (and smaller) time unit the time taken by c_1 to pass *one* of the B units.

Although Zeno could hardly have grasped the full mathematical implication of his arguments, they served to worry the contemporary mathematicians – as did the discovery of the irrational discussed in Chapter 3 – and they made them aware of the subtlety of reasoning about the infinite. The paradox of the Stade employs what has remained one of the most powerful methods of reasoning in mathematics ('reductio ad absurdum'), in which the truth of a proposition is postulated and it is shown by logical argument that the postulate eventually leads to a contradiction.

We turn to the problem of areas, in particular the problem of 'squaring the circle'. This was one of the 'three famous mathematical problems of antiquity': (i) *the trisection of the angle*: to divide a given angle into three equal parts; (ii) *the duplication of the cube*: to find the side of a cube whose volume is twice that of a given cube; (iii) *the quadrature of the circle*: to find a square of area equal to that of a given circle. The second problem is the so-called 'Delian' problem: the story goes that an Athenian delegation, sent to ask the oracle of Apollo at Delos how to avert the plague, was told that 'the cubical altar to Apollo must be doubled'. However, construction of an altar with a side twice that of the old one did not cure the plague! These problems are

Fig. 8.4

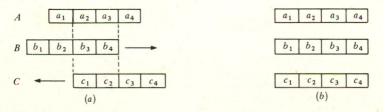

(a) (b)

important in the history of mathematics because they cannot be solved, except approximately, by construction of a finite number of straight lines and circles, i.e. solution is impossible by ruler-and-compass construction.† In their efforts to deal with this challenge mathematicians were led to try new methods which opened up new fields of mathematics. The Greeks were led to the discovery of the conic sections and of higher-order curves such as the Archimedean spiral (Boyer 1968, p. 140); and in modern times the ancient Greek problems have found their full resolution within the context of Galois theory and transcendental number theory (see pp. 64, 68), quite abstract branches of mathematics.

An early contribution to the problem of the quadrature of the circle was made by Hippocrates of Chios (a contemporary of Zeno). It is the only complete mathematical fragment surviving from the Golden Age of Greece, and is interesting because it indicates that Euclid's strict axiomatic approach to geometrical problems existed more than a hundred years before Euclid's time. The fragment deals with a typically useless, but theoretically interesting, problem: that of determining the areas of circular 'lunes' which are crescents bounded by two circular arcs. Hippocrates shows the following (Fig. 8.5): suppose circle 1 has centre M and diameter AB. Let MC be a radius perpendicular to AB and draw circle 2 with centre C passing through A and B. This defines the shaded lune: Hippocrates proves that its area is equal to that of the square on the radius MB. The proof assumes as known the result that the areas of two circles are in the same ratio as the squares on their radii (and hence also that two circular segments, like the shaded areas in Fig. 8.6, which subtend the same angle at the centre, are in the ratio of the squares on the radii). The Greeks had a rigorous method for proving such results, the so-called 'method of exhaustion'

---

† According to Plato the only perfect geometrical figures are straight lines and circles; hence the preference in Greek geometry for the use of ruler and compass only (where the 'ruler' is an unmarked straight edge). Many constructions are possible by these methods: thus a line can be divided into an arbitrary number of equal parts, any angle can be bisected, a square can be constructed equal in area to a given polygon, and so on.

which we describe below; but its discovery is usually ascribed to Eudoxus who lived later than Hippocrates (around 370 B.C.), and we do not know how Hippocrates proved this result.

We subdivide the lune by drawing two tangents to circle 2, at A and at B (Fig. 8.7). Then α is a segment of circle 1 subtending a right angle at M, and δ is a segment of circle 2 subtending a right angle at C. Hence

$$\alpha : \delta = AM^2 : AC^2 = 1 : 2 \text{ (by the theorem of Pythagoras)},$$

and clearly $\alpha = \beta$; hence $\alpha + \beta = \delta$, and thus the area of the lune, which is $\alpha + \beta + \gamma$, is equal to $\delta + \gamma$, i.e. the area of the triangle ABD, and this is BM^2 as required.

Hippocrates's result is important, for it is the first demonstration in the history of mathematics that a curvilinear area *can* be commensurable with a straight-sided figure. Surely this should also be possible for the simplest curvilinear area, the full circle. It seems natural to suppose that Hippocrates must have tried to divide the whole circle into lunes which could all be 'squared' as above, and in this way to achieve the quadrature of the circle. This seems a very natural thing to try; unfortunately

Fig. 8.5

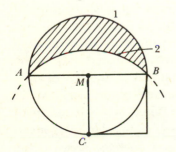

Fig. 8.6

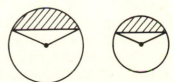

Fig. 8.7

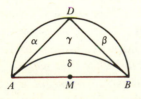

the trick does not work for the whole circle! As we noted on p. 67, the final proof that this was not possible was not given until more than 2000 years later.

The *Sophists* were a group of professional teachers in Athens, active towards the end of the fifth century B.C. They were quite unlike the Pythagoreans: while the latter were forbidden to accept payment for their teaching, the Sophists supported themselves by tutoring, not only in honest study, but also in the art of 'making the worse appear the better'. (We have to remember that our descriptions of the Sophists are due to Plato, Socrates and others who opposed their philosophy in general and who, when they described a Sophist as 'vain, boastful and acquisitive', were unlikely to be unprejudiced observers.) One of the Sophists, Antiphon, claimed to have squared the circle by the method, mentioned at the beginning of this chapter, of inscribing a regular polygon, doubling the number of sides, and continuing this process until the polygon became indistinguishable from the circle. While they realised that one can, in this way, *estimate* the area of a circle with an accuracy sufficient for all practical purposes, the Greek mathematicians also saw quite clearly that a straight-sided figure, however large the number of sides, was in principle different from the ideal curvilinear circle, and they did not accept Antiphon's argument as a valid solution of the problem. Antiphon's suggestion was nevertheless important as it provided the germ of Eudoxus's 'method of exhaustion' (the name itself originated only in the seventeenth century). We shall illustrate the method by using it to prove the result used above, that the areas of two circles are in the ratio of the squares on their radii. What is the real nature of the problem? Suppose we inscribe n-sided regular polygons in the two circles; then it is easy to show that the areas of the *polygons* are as the squares on the radii of the circles however large the value of n (as we noted on p. 86, there is no difficulty with areas of straight-sided figures). What we must *prove* is that the difference in area between the n-sided polygon and the circle (which is always non-zero however large n is) does not affect this result. Nowadays we would say that we are dealing with a *limit*. The method of exhaustion provides a logically precise method of argument which avoids the difficulties associated with arguing about 'infinitely many' 'infinitely small' quantities. It is in fact a primitive form of integral calculus.

The method rests on the 'axiom of continuity' (given in Euclid, and sometimes called the 'axiom of Archimedes'), which says essentially that, if A and B are magnitudes of the same kind, and A is less than B, then one can always find a multiple of A which is greater than B. (This seems 'obvious' and is certainly valid for finite non-zero real numbers. It excludes from consideration vaguely conceived 'infinitesimally small' quantities. Nowadays (see also Chapter 10) we know how to construct 'non-Archimedean' fields which allow us, if we wish, to give precise definitions of both 'infinitely large' and 'infinitely small' numbers.) From the axiom of continuity the Greeks deduced the equivalent proposition: if α is greater than ε, and if one subtracts from α at least one-half of α, then from the rest at least one-half again, and so on, one will eventually be left with a magnitude which is less than ε. (*Proof*: given ε, we can find n such that $n\varepsilon > \alpha$. Now $2\varepsilon = 2\varepsilon$, $3\varepsilon < 2^2\varepsilon$, $4\varepsilon < 2^3\varepsilon$, and so on; finally $\alpha < n\varepsilon < 2^{n-1}\varepsilon$. Thus we can go on *doubling* ε until it is greater than α; or, equivalently, we can go on *halving* α until it is less than ε.) Now we can formulate our proof. We start the process of 'exhausting' the area of the circle with straight-sided figures by inscribing a square $ABCD$ in the circle (Fig. 8.8); this

Fig. 8.8

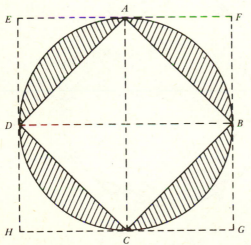

differs from the circle by the shaded region (S). Add to the figure
four triangles ABF, BCG, CDH, DAE as shown; their total area
equals the area of the inscribed square. The area S is less than
the area of the four triangles, hence S is less than the area of the
inscribed square, which equals the area of the circle minus S;
thus we see that S is less than *half* the area of the circle. We
continue by removing four triangular areas ABW, BCX, CDY,
DAZ from S (Fig. 8.9), giving us an inscribed *octagon*
$AWBXCYDZ$. The four triangles clearly have half the area of the
four rectangular boxes such as $ABPQ$. We see from the figure
that the (shaded) difference in area (S') between the octagon and
the circle is less than the sum of the areas of the four triangles.
Thus we have shown that, if one cuts from the circle the
inscribed square, one removes more than half the area; if one
now cuts out the four triangles (thus removing an octagon), one
again removes more than half of the remaining area. And so the
exhaustion process continues; the next step is to erect eight
triangles on the sides of the octagon to give a 16-gon, and this
removes more than half the area S'.

Fig. 8.9

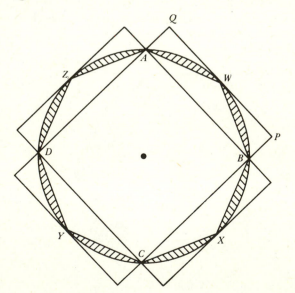

How does one clinch the argument? At this stage the axiom of continuity is used. Suppose ε is an arbitrary area greater than zero. Eventually (see above) we must get to a stage where the difference in area between the inscribed polygon and the circle is less than ε. Suppose the circles we wish to compare have areas A, A' and radii r, r'; we want to prove that $A : A' = r^2 : r'^2$. Suppose B is the area of a polygon inscribed in A for which $A - B < \varepsilon$, (or $A - \varepsilon < B$). Now suppose the result is wrong because A is 'too big'; then (for *some* positive δ) we must have

$$\frac{A - \delta}{A'} = \frac{r^2}{r'^2}.$$

Choose ε (which is arbitrary) to be not greater than δ, then $A - \delta < B$. Suppose B' is the area of the corresponding polygon in A', so that $B' < A'$, then we have the inequalities

$$B > A - \delta, \qquad B' < A'.$$

Hence

$\dfrac{B}{B'}$ is certainly greater than $\dfrac{A - \delta}{A'} = \dfrac{r^2}{r'^2}.$

But we *know* that $B/B' = r^2/r'^2$ (since our result is known to be true for polygons), and thus we have a contradiction. In the same way (try this for yourself) we can dispose of the assumption that A is 'too small', and our result is thus proved. This is a logically rigorous proof (given by Euclid): all the mystery of the infinite process lies in the axiom of continuity!

Of the ancients it was Archimedes who made the most beautiful use of the method of exhaustion and who came nearest to integration in the modern sense. (For fuller details on Archimedes's works, see Dijksterhuis 1957.) Archimedes, who lived approximately 287–212 B.C., was the leading mathematician of the Hellenistic age. He lived and died in Syracuse in Sicily. Many picturesque legends about him are related by the Roman historians, but we have little hard fact concerning his life. He was a great pure mathematician who also dealt with practical matters; he was an inventor of ingenious war machines and one of the founders of the science of mechanics. Archimedes is an early example of the conflicts that arise between 'pure' and 'applied' mathematics. According to Plutarch: 'he regarded as sordid and ignoble the construction of instruments, and in general every art directed to use and profit, and he only strove after those things which, in their beauty and excellence, remain beyond all contact with the common needs of life'. So he did pure mathematics: 'continually bewitched by a siren who always accompanied him, he forgot to nourish himself and omitted to care for his body; and when, as would often happen, he was urged by force to bathe and anoint himself, he would still be drawing geometrical figures in the ashes or with his finger would draw lines on his anointed body, being possessed by a great ecstasy and in truth a thrall to the Muses'. But (as happens these days too) he was told to make himself useful: 'most of [his technical inventions] were the diversions of a geometry at play which he had practised formerly, when King Hieron had emphatically requested and persuaded him to direct his art a little away from the abstract and towards the concrete, and to reveal his mind to the ordinary man by occupying himself in some tangible manner with the demands of reality'.

The story of how the works of Archimedes have come down to us is interesting; in order to learn how the historian of science actually gets his material, read Chapter 2 of Dijksterhuis (1957). We are fortunate that a large part of Archimedes's extensive writings have been preserved, but we also know from these writings that there are other works (in particular one on the centres of gravity of solids) which are lost.

Archimedes's results include the formulation of the 'law of the lever' in statics and of 'Archimedes's principle' in hydrostatics. He was fascinated by very large numbers, and in *The Sand-Reckoner* he estimated the number of grains of sand which would fill the universe. (His answer, about 10^{63} in our notation, is not too far from modern estimates of the number of atoms in the universe.) The 'spiral of Archimedes' is defined as the plane locus of a point which moves uniformly outwards along a line from the origin, while the line itself rotates uniformly about the origin (the polar equation is $r = k\theta$). With this curve, as Archimedes showed, the trisection of the angle and the quadrature of the circle are easily accomplished (see Boyer 1968, p. 141).

Archimedes also obtained an excellent estimate of π. Given a circle, he calculated the perimeter of inscribed and circumscribed regular polygons (Fig. 8.10), thus obtaining upper and lower bounds for the circumference. Archimedes gives a rule for going systematically from any polygon with n sides to one with $2n$ sides (this only requires bisection of angles); thus he had a systematic method of calculating the irrational number π to any desired accuracy. Note that this was not an attempt to 'square the circle' in the sense of Hippocrates, but to find (in modern terms) a good 'rational approximation' for the ratio of the circumference of a circle to its diameter. The details of the calculation are elaborate and cunning. Archimedes goes as far as a polygon with 96 sides, and he needs rational approximations for square roots like $\sqrt{3}$. He takes, in fact, $\sqrt{3} \approx 1351/780$, a very close estimate $(\sqrt{3} = 1.732\,050\,8\ldots, \quad 1351/780 = 1.732\,051\,2\ldots)$, but he does not say how he got this result, and

Fig. 8.10

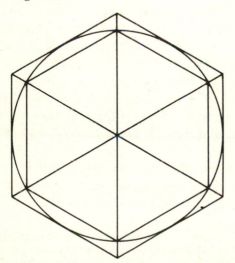

there has been much speculation on this question.† Archimedes's final result, in modern notation, is that $3\frac{10}{71} < \pi < 3\frac{1}{7}$.

Our chief interest, from the point of view of the early calculus, is in Archimedes's famous *quadrature of the parabola*. This was a sensational achievement: the conic sections (ellipse, hyperbola, parabola) were quite well known at the time of Archimedes, and the quadrature of circle, ellipse and hyperbola had all been tried by many Greek mathematicians, always in vain; then the young Archimedes had the idea to try the parabola and was successful! Archimedes in fact gives two different proofs, both of which contain ideas which were taken up again much later. We therefore outline both of them. We define the parabola (Fig. 8.11) as the locus of a point C such that $BC/EF = BO^2/EO^2$ (this is equivalent to the modern equation $y = x^2$), and we want to *prove* that the area of the segment FOF' is two-thirds the area of the rectangle $FF'E'E$, thus showing that the parabolic segment is commensurable with a rectangle.

First proof. We 'exhaust' the parabolic area by means of triangles (Fig. 8.12). The first triangle is FOF'; we denote its area by $\triangle FOF'$. The second stage is to add the triangles FOC, $F'OC'$, where C, C' are defined by dividing the line $E'OE$ at B, B' such that $BO = \frac{1}{2}EO$, $B'O = \frac{1}{2}E'O$, and drawing vertical lines through B, B' to meet the parabola at C, C'. Since $BO = \frac{1}{2}EO$, by similar triangles (see Fig. 8.12), $BG = \frac{1}{2}BD$. We shall show that $\triangle FOC + \triangle F'OC' = \frac{1}{4}\triangle FOF'$. From the definition of the parabola we have $BC = \frac{1}{4}BD$, so $DG = 2GC$. Thus the triangles DGF, GCF have the same height and have bases in the ratio $2:1$. Hence $\triangle DGF = 2\triangle GCF$, and similarly $\triangle DGO = 2\triangle GCO$. Adding, we have $\triangle FOD = 2\triangle FOC$, Hence $\triangle FOA = 2\triangle FOD = 4\triangle FOC$, and thus $\triangle FOC + \triangle F'OC' = \frac{1}{4}\triangle FOF'$, as required.

We now continue this process, adding four further triangles as indicated in Fig. 8.13, and we find similarly that $\triangle FCC_1 + \triangle OCC_2 = \frac{1}{4}\triangle FOC$, so (for both sides) we have that the area of

† Because of the use of a primitive number system, Archimedes had trouble with handling complicated fractions. But (see p. 3) the Babylonians, already about 2000 B.C., knew how to calculate square roots with considerable accuracy.

Fig. 8.11

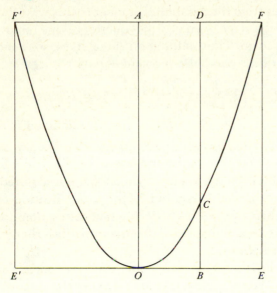

Fig. 8.12

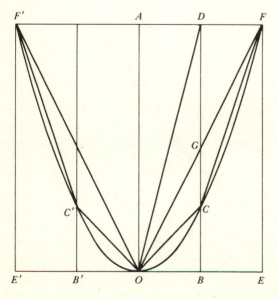

the added triangles equals $(1/4^2)\triangle FOF'$. Removing more and more triangles from the parabolic segment in this way, the area removed at each stage is exactly one-quarter of that removed at the previous stage. Thus, adding all these areas, we conclude that the area of the parabolic segment equals

$$\triangle FOF' \times \left[1 + \frac{1}{4} + \frac{1}{4^2} + \frac{1}{4^3} + \cdots \right] = \tfrac{4}{3}\triangle FOF'$$

$$= \tfrac{2}{3} \times \text{rectangle } FF'E'E,$$

as was to be proved. Note that Archimedes did not say, at this final stage, 'we now sum an infinite series'; that had no rigorous meaning for him. But he knew very well that the required sum was $\tfrac{4}{3}$, and he proved it rigorously by an application of the method of exhaustion, i.e. by showing that the assumptions (i) that the sum is greater than $\tfrac{4}{3}$, (ii) that the sum is less than $\tfrac{4}{3}$, both lead to contradictions.

Second proof (this is close to modern integration, as defined by Riemann). We wish to find the area R *under* the parabolic arc lying above OE, where we take $OE = 1$ (Fig. 8.14). We divide

Fig. 8.13

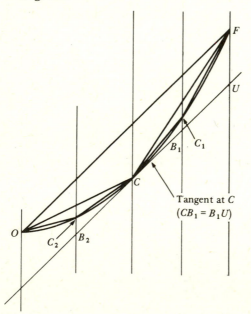

OE into four equal parts and erect rectangles of width $\frac{1}{4}$ as shown. The area R then clearly lies between S_4 and T_4, where the 'lower sum' S_4 is the sum of the areas of the rectangles lying under the parabola,

$$S_4 = \frac{1}{4}0^2 + \frac{1}{4}\left(\frac{1}{4}\right)^2 + \frac{1}{4}\left(\frac{2}{4}\right)^2 + \frac{1}{4}\left(\frac{3}{4}\right)^2,$$

and the 'upper sum' is clearly

$$T_4 = \frac{1}{4}\left(\frac{1}{4}\right)^2 + \frac{1}{4}\left(\frac{2}{4}\right)^2 + \frac{1}{4}\left(\frac{3}{4}\right)^2 + \frac{1}{4}\left(\frac{4}{4}\right)^2.$$

If we now divide the interval OE into n equal parts, then we have, similarly, $S_n < R < T_n$, with

$$S_n = \frac{1}{n}0^2 + \frac{1}{n}\left(\frac{1}{n}\right)^2 + \frac{1}{n}\left(\frac{2}{n}\right)^2 + \cdots + \frac{1}{n}\left(\frac{n-1}{n}\right)^2,$$

and

$$T_n = \frac{1}{n}\left(\frac{1}{n}\right)^2 + \frac{1}{n}\left(\frac{2}{n}\right)^2 + \cdots + \frac{1}{n}\left(\frac{n}{n}\right)^2.$$

This holds for any n. We see that the difference $T_n - S_n = (1/n)\,(n/n)^2$ which tends to zero as n increases. Also $T_n = (1/n^3)(1^2 + 2^2 + \cdots + n^2)$. Thus we do not now have a geometric series to sum, but Archimedes knew that $1^2 + 2^2 + \cdots + n^2 = \frac{1}{6}n(n+1)(2n+1)$. Hence

$$T_n = \frac{1}{6}\left(1 + \frac{1}{n}\right)\left(2 + \frac{1}{n}\right) \to \frac{1}{3} \quad \text{as } n \to \infty.$$

Fig. 8.14

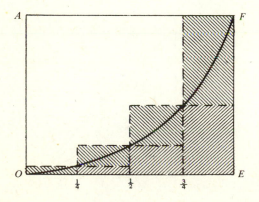

(Again the limits were established using the method of exhaustion.) Thus we have shown that the area R under the parabola is one-third the area of the (unit) square $OEFA$. The area of the parabolic segement OFA therefore equals two-thirds the area of the square $OEFA$, which is our previous result.

Archimedes could calculate volumes and centres of gravity of many solids. How did he get his results? Most of his treatises are highly rigorous and logically precise, like modern papers in pure mathematics, and give little hint of the actual way the results were obtained. And in fact the method of exhaustion, although rigorous, is essentially a way of proving a result once it is already known, since one must show that any assumption *other* than the known result leads to a contradiction; thus the method is not a technique for discovering new and unexpected results.

We now know something of Archimedes's way of thought through a remarkable discovery, made as recently as 1906, of a previously unknown work by Archimedes. J. L. Heiberg, a Danish historian of science, heard of the existence in Constantinople of a palimpsest with mathematical content (a palimpsest is a parchment where the original text has been covered by a different one). He found that, underneath a collection of prayers for the Eastern Orthodox Church, there was a mathematical text which had been copied in the tenth century. This text consisted of several well-known works by Archimedes, together with a new one called simply *The Method*. It was in the form of a letter

Fig. 8.15

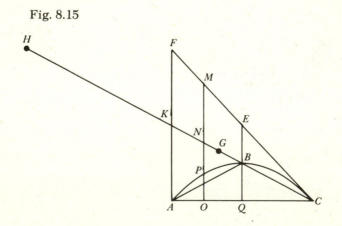

to Eratosthenes, a mathematician and librarian at the University of Alexandria (remembered for originating the so-called 'sieve method' for obtaining prime numbers). The letter describes a non-rigorous 'mechanical' approach which Archimedes used as his guide to discovery. It involves an 'atomistic' way of thinking, in which areas are thought of as the totality of a very large number of constituent lines, and volumes as the totality of the constituent areas. (This is similar to the way a 'naive' scientist nowadays tends to think when he applies calculus methods.) But Archimedes, in his refusal to use this approach in his published works, was well aware of its tentative nature (how many lines are needed, and are they of zero, or 'very small', thickness?).

We illustrate the ingenious use made by Archimedes of the 'Method' by deriving (once more) the result on the area of a parabolic segment (Fig. 8.15). Archimedes thought of the segment $ABCA$ as the sum of all parallel lines such as OP, and he wanted to compare this area with the area of the triangle AFC (FC is the tangent to the parabola at C). To do this he balanced the constituent lines like weights in mechanics, using the 'law of the lever'. In Fig. 8.15, CK is a median of the triangle AFC, and $HK = KC$. Archimedes placed a weight equal to the length of the line OP at H, and noted that this exactly balances the line OM placed at its centre of gravity N, with the point K as fulcrum.† If we do this balancing for *all* parallel lines such as OP, then we see that the whole parabolic area placed at H balances the whole triangle AFC, and thus balances the area of the triangle placed at its centre of gravity G, where $KG = \frac{1}{3}KC$. Hence the parabolic area must be $\frac{1}{3}\triangle AFC = \frac{4}{3}\triangle ABC$, the previous result. Boyer (1968) says about the discovery of the 'Method': 'In a sense the palimpsest is symbolic of the contribution of the Medieval Age. Intense preoccupation with religious concerns nearly wiped out one of the most important works of the greatest mathematician of antiquity; yet in the end it was medieval scholarship that inadvertently preserved this, and much besides, which might otherwise have been lost.'

† From the geometry of the parabola we have $OM/OP = AC/AO$ (check this); hence $OM/OP = KC/KN = HK/KN$, i.e. $OP.HK = OM.KN$, as required by the law of the lever.

9

The calculus in the seventeenth century

After Archimedes 1900 years elapsed before further significant advances were made in the calculus, but then things happened fast. The hundred-year span of the seventeenth century was a 'heroic age' during which the infinitesimal calculus grew from rudimentary beginnings into a highly developed mathematical discipline. The discovery of the calculus is usually attributed to Newton and Leibniz, and associated with the period 1665–75. There is no reason to dispute this attribution, but it would be wrong to imagine that these two men invented the calculus more or less 'out of the blue'. On the contrary, 'infinitesimal methods' were very much 'in the air' in their time, and virtually every mathematician of note in the seventeenth century contributed to their development. In this period it makes even less sense than usual to argue about who discovered what, and when a result is associated with a particular author all one can really say with any certainty is that the result was known to that author at the date in question.

Why should such a tremendous advance have occurred in the seventeenth century, after such a long period of stagnation? It was undoubtedly part of the new spirit in art and science associated with the Renaissance. Editions of the works of Archimedes and other Greek mathematicians had been published during the sixteenth century, so the Greek methods for finding areas, volumes and centres of gravity were known to scholars. Strong stimuli for the development of theoretical mechanics, and thus for the elaboration of computational methods and the application of infinitesimal methods to the study of motion and change, came from sources such as the increasing use of machines in early forms of industry, for example pumps and lifts in mining; the 'new astronomy' asso-

ciated with Copernicus, Tycho Brahe and Kepler which gave
hope that the science of mechanics might account for celestial as
well as earthly phenomena; the perfection of clocks leading to
accurate measurement of time and demonstrating the occurrence of order and regularity in natural events.

Characteristic of all the new work on the calculus was the
abandonment of Archimedean standards of rigour. People
wanted to get results, and they sensed that the methods under
study could be powerful aids for making new discoveries; they
therefore pushed ahead without worrying too much about logical deficiencies in the argument, though they usually realised
quite well that the rigorous foundation was lacking. Each
author tended to set his own standard of rigour. This attitude
persisted throughout the seventeenth and eighteenth centuries,
and it was only after 1800 that Greek standards were re-
introduced to put the whole subject on a logically sound
basis, turning it from 'calculus' into 'analysis'.

We have space to mention only a few of the more important
contributions to the calculus in the period immediately preced-
ing the time of Newton and Leibniz. Cavalieri, a disciple of
Galileo and professor at Bologna, published in 1635 his
Geometria Indivisibilibus Continuorum, a systematic account
of infinitesimal methods which did much to stimulate interest
in such problems. Cavalieri's calculus is related to Archimedes's *Method* with which Cavalieri was, however, presumably not acquainted. It regards a plane area as made up of lines
('indivisibles'), and a solid volume as composed of areas. The
'indivisibles' were not so much the small but finite 'atoms' of the
Greeks, but were derived from medieval scholastic philosophy
which considered every continuum to be indefinitely subdivisible. A typical result obtained by Cavalieri is the following: consider the areas F, G under two curves f, g (Fig. 9.1).
Draw the curve h which is such that the ordinate h' of any point
is the sum of the corresponding ordinates of points on f and g:
$h' = f' + g'$. The method of indivisibles deduces from this that
the sum of all ordinates f' (i.e. the area F) plus the sum of all
ordinates g' equals the sum of all ordinates h'; hence the area H
under the curve h equals $F + G$. This conclusion is correct, but it

was soon realised that this type of argument can also give results which are obviously wrong: applied to two circles regarded as the sums of all their radii one would conclude (Fig. 9.2) that, if $OQ = 2OP$, then the area of the larger circle must be twice that of the smaller, and this of course is incorrect. Again, suppose we divide a triangle into two parts A and B as shown in Fig. 9.3. With each constituent line of part A we can associate an equal line of part B and *vice versa*, as shown; thus the area of A should be equal to the area of B (this difficulty was pointed out by Torricelli, another disciple of Galileo). Thus it was clear that the method of indivisibles, in Cavalieri's form, could not provide a satisfactory general definition of areas and volumes.

Cavalieri's most enduring contribution was his generalisation of Archimedes's quadrature of the parabola. He considered

Fig. 9.1

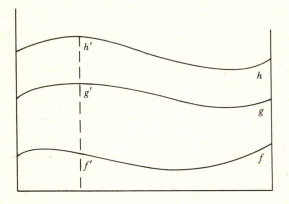

Fig. 9.2

Fig. 9.3

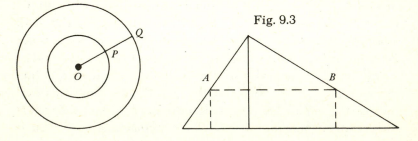

the area under the cubic curve $y = x^3$ between $x = 0$ and $x = 1$ and obtained, summing rectangular areas in the manner of Archimedes's second proof (p. 101).

$$T_n = \frac{1}{n^4}(1^3 + 2^3 + \cdots + n^3).$$

He thus needed to be able to sum the first n *cubes* of the integers. A closed formula for this sum had been known from ancient times. The Arabs had the following elegant construction: take a square of side $1 + 2 + 3 + \cdots + n$ and subdivide it in the manner shown in Fig. 9.4. Since

$$1 + 2 + 3 + \cdots + n = \tfrac{1}{2}n(n+1),$$

the area of the square is $[\tfrac{1}{2}n(n+1)]^2$. But we can also evaluate the area as the sum of the areas of the L-shaped pieces: the area of the nth piece consists of two rectangles and is clearly

$$n[\tfrac{1}{2}n(n+1)] + n[\tfrac{1}{2}(n-1)n] = n(\tfrac{1}{2}n^2 + \tfrac{1}{2}n + \tfrac{1}{2}n^2 - \tfrac{1}{2}n) = n^3.$$

Hence

$$1^3 + 2^3 + \cdots + n^3 = [\tfrac{1}{2}n(n+1)]^2,$$

and therefore

$$T_n = \frac{1}{4}\frac{n^2(n+1)^2}{n^4} = \frac{1}{4}\left(1 + \frac{1}{n}\right)^2 \to \frac{1}{4} \quad \text{as } n \to \infty.$$

As in the case of the parabola the result is a rational number, so

Fig. 9.4

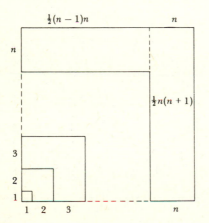

the area in question is commensurable with a rectangle. Cavalieri went on to work out the corresponding area for $y = x^4$ (result $\frac{1}{5}$), $y = x^5$ (result $\frac{1}{6}$), and so on. Each case had to be considered separately, and the labour involved increased at each step! When he had reached $y = x^9$ (with result $\frac{1}{10}$) Cavalieri gave up, and he could not see how to deal with the case $y = x^k$ for general integral k. The difficulty was overcome by Fermat, around 1650. He realised that in order to approximate the area with rectangles it was not necessary to divide the interval $(0, 1)$ into *equal* parts, in the manner of Archimedes's second proof, and that a clever choice of subdivision could lead to a simpler summation. This was a brilliant idea, fundamental for the further development of the integral calculus; it was really a new form of Archimedes's *first* proof for the parabola. Fermat's subdivision is shown in Fig. 9.5; ρ is some number between 0 and 1. The total area of the rectangles under the curve is the infinite sum

$$S_\rho = (1-\rho)\rho^k + (\rho - \rho^2)(\rho^2)^k + (\rho^2 - \rho^3)(\rho^3)^k + \cdots$$
$$= (1-\rho)\rho^k + (1-\rho)\rho^{2k+1} + (1-\rho)\rho^{3k+2} + \cdots$$
$$= (1-\rho)\rho^k[1 + \rho^{k+1} + \rho^{2k+2} + \cdots].$$

Summing the infinite geometric series we have

$$S_\rho = (1-\rho)\rho^k/(1-\rho^{k+1}),$$

Fig. 9.5

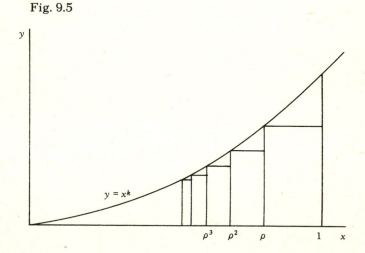

$y = x^k$

ρ^3 ρ^2 ρ 1 x

and, since

$$1-\rho^{k+1}=(1-\rho)(1+\rho+\rho^2+\cdots+\rho^k),$$
$$S_\rho=\rho^k/(1+\rho+\rho^2+\cdots+\rho^k).$$

At this stage we let ρ approach 1; the width of each of the rectangles then tends to zero, and their total area tends to the area R under the curve. Thus $R=S_{\rho=1}=1/(k+1)$, the general result suggested by Cavalieri's calculations. Fermat noted that this argument is easily extended to the case when k is a fraction, say $k=p/q$ with p,q positive integers. It is in fact only in the last step, the factorisation of $1-\rho^{k+1}$, that we assumed k to be integral; we now obtain instead (with $t^q=\rho$)

$$S_\rho=\frac{(1-\rho)\rho^{p/q}}{1-\rho^{(p+q)/q}}=\frac{(1-t^q)t^p}{1-t^{p+q}}$$
$$=\frac{t^p(1-t)(1+t+t^2+\cdots+t^{q-1})}{(1-t)(1+t+t^2+\cdots+t^{p+q-1})}.$$

As ρ and t approach 1, this tends to

$$\frac{q}{p+q}=\frac{1}{(p/q)+1}=\frac{1}{k+1}$$

as before. Negative k can also be dealt with, and Fermat was thus able to integrate x^k for all rational k except $k=-1$; for this special value the method does not give a geometric series and fails. This special case is the problem of the area under the *hyperbola, $y=1/x$.*

A method of subdivision for finding the area under this curve was, it seems, first given by a Belgian Jesuit scholar, Gregory of Saint-Vincent, in 1647. It appears at the end of an enormous and useless work on squaring the circle and brought its author little credit; presumably he was not aware of the importance of his discovery. Gregory's idea is to fill the area under the curve $y=1/x$ between (say) $x=1$ and $x=2$ with a certain number of rectangles (Fig. 9.6); then to fill the area between 2 and 4 with the *same* number of rectangles (which therefore have double the width); then the area between 4 and 8 with the same number again (again doubling the width); and so on. Since $y=1/x$, *doubling* the value of x means *halving* the value of y; hence a rectangle of doubled width also has its height halved, so that its area is unchanged. Thus the total area of the rectangles between $x = 1$ and $x = 2$ is equal to the total area of the rectangles between $x = 2$ and $x = 4$, and so on. This conclusion remains unchanged if we increase the number of rectangles used in each segment of the x-axis, and so it follows, on increasing the number indefinitely, that the property of the rectangles must also hold for the *area $J_{a,b}$ under the curve* between $x = a$ and $x = b$: $J_{1,2} = J_{2,4} = J_{4,8} = J_{3,6} = \cdots$. The general conclusion is that $J_{a,b} = J_{ka,kb}$, where k is any positive integer, and it is easy to generalise to the case where k is any positive fraction (try to show this yourself). It follows that the area under the curve between 1 and xy is

$$J_{1,xy} = J_{1,x} + J_{x,xy} \text{ (since areas add)} = J_{1,x} + J_{1,y}.$$

This shows that the *area function $J_{1,x}$* has the basic property of a *logarithm*, $\log xy = \log x + \log y$.

Does this mean that Gregory *discovered* the formula $\int dx/x = \log x$ which we find nowadays in tables of integrals? We have already suggested that questions like this can hardly be answered; there is no single discoverer. We do know that the formula in its familiar form was given by Leibniz. We must be

Fig. 9.6

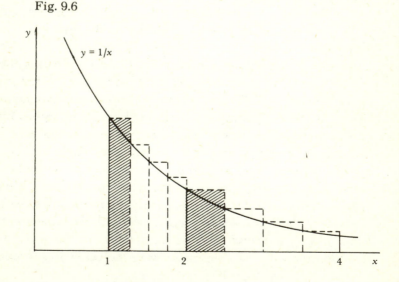

careful not to attribute our modern way of thinking to the people who were groping their way towards the calculus in the early seventeenth century. Nowadays the definition of an integral is given in analysis as the limit of a certain sum, and intuitive geometrical concepts play no part in the definition. We can then *define* the area under the curve $y = 1/x$ as a certain definite integral. Gregory on the other hand was concerned with the *geometrical* problem of areas; others at his time had become familiar with logarithms as *numbers* which were useful aids to calculation (see Chapter 1). The two concepts were quite distinct, and $\log x$ was not yet regarded as a *function* worth studying for its mathematical significance. What Gregory showed – and it was a key step – was that the area under the hyperbola between a fixed and a variable abscissa *behaves like a logarithm*, increasing 'arithmetically' as the abscissa increases 'geometrically'.

Thus by about 1650 a fair amount of progress had been made with the ancient *quadrature problem* of finding areas under curves. It was also found that related geometrical problems, such as the 'rectification' of various curves (i.e. finding a straight line of the same *length*), and determining areas of surfaces of revolution, could all be reduced to quadratures. At first problems were classified in accordance with the geometrical (or mechanical) situation from which they arose, but it was gradually realised that different problems, whatever their origin, were mathematically equivalent if they led to the same quadrature problem; in this way the integral calculus came into being as a separate mathematical discipline in its own right.

How about the other basic concept of the calculus, the *derivative*? It is implicit in several classes of problems dating back to antiquity: the geometrical problem of finding the directions of *tangents* to curves; the *maximum/minimum* problem of finding the greatest and least values of variable magnitudes; and the kinematical problem of the *speed* of a moving object, i.e. the rate at which its position changes with time. Progress with all these questions contributed to the creation of the differential calculus in the seventeenth century. The Greeks had no differential calculus analogous to the method of exhaustion for finding areas. They determined tangents to simple curves by geometrical construction: for the circle as the line perpendicular to the radius, and for the ellipse as the line making equal angles with the lines to the foci (Fig. 9.7). (Note that for both these curves the Greeks found the *area problem* insoluble!) Indeed the properties of tangents and normals to all the conic sections had been thoroughly studied in Apollonius's famous *Conics*. But the Greeks had no satisfactory general definition of the tangent to an *arbitrary* curve; they had only the vague notion of it as a line such that no other line could be drawn through the point of contact to lie 'between' the tangent and the curve. The only hint of differential-calculus ideas comes in an observation of Archimedes who determined the direction of the tangent to his spiral $r = k\theta$ (see p. 96) by thinking in terms of the instantaneous direction of motion of a moving point, the motion being the resultant of a uniform radial motion away from the origin and a circular motion about the origin. This was however an isolated result.

It seems that Fermat, around 1630, first clearly knew how to differentiate a simple function. (Pierre de Fermat (1601–65) was a lawyer who did mathematics as a hobby and who published very little during his lifetime. Thus the exact date of his many profound and original discoveries is uncertain, and he did not during his life receive the credit he deserved. His deep contributions to the theory of numbers were mentioned in Chapter 2. He developed analytic geometry independently of Descartes who is usually regarded as its founder, and Fermat was also, with Blaise Pascal, a founder of the mathematical theory of probability. The introduction of powerful algebraic methods

Fig. 9.7

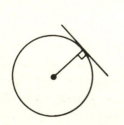
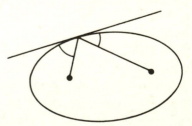

into geometry was itself one of the advances needed for the full development of the calculus.)

To find the slope of the tangent at say $x = 1$ to the curve $y = x^k$, Fermat used an argument equivalent to differentiation (Fig. 9.8). He regarded the direction of the tangent at P as given by the limiting position of the chord PP' when P' approaches P, so that the slope of the tangent is the limiting value of $P'Q/PQ$, i.e.

$$\lim_{x \to 1} \frac{x^k - 1}{x - 1}$$

which Fermat could evaluate as k. We see that in the determination of this limit Fermat had to deal with essentially the same mathematical problem as in his calculation of the area under the same curve. Was Fermat aware of the relation between differentiation and integration? He does not mention it, and at first the techniques of differentiation were developed quite separately from those of integration – they arose, after all, from problems that appeared to have no obvious relation to each other.

At some stage in the seventeenth century it was realised – we cannot name any single discoverer – that differentiation and integration were inverse mathematical processes; a realisation that was to play a fundamental part in the subsequent development of the calculus as a systematic discipline. This inverse relation involves the idea of the *indefinite integral*: if $f(x)$ is a given function and $F(x)$ is any function whose derivative $F'(x)$ equals $f(x)$, we call $F(x)$ an indefinite integral (or

Fig. 9.8

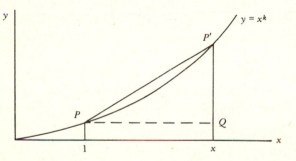

antiderivative) of $f(x)$ and write it $\int f(x)\,dx$ (it is not unique: if $F(x)$ is an indefinite integral of $f(x)$ so is $F(x)+c$, where c is any constant).

What has this notion to do with areas under curves? It is necessary to *show* that, if $f(x)$ has an indefinite integral $F(x)$, the area under the curve $y = f(x)$ between $x = a$ and $x = b$ is $F(b) - F(a)$ (with certain provisos regarding signs. This is of course the standard rule for evaluating 'definite integrals' $\int_a^b f(x)\,dx$ in terms of indefinite integrals). Newton gave such a proof for the curve $y = ax^{m/n}$, but several of his predecessors were undoubtedly aware of the idea. (All these proofs relied on 'area' as a geometrically plausible idea which needs no further definition.) If we now regard the 'upper limit' b as a variable quantity, the inverse relation between differentiation and integration can be expressed by the two equations

$$F(b) = \int_a^b f(x)\,dx + \text{constant}, \qquad F'(b) = f(b).$$

The approach to quadratures via the indefinite integral, adopted by Newton and Leibniz and their successors, elevates *differentiation* into the primary concept and makes *integration* a secondary idea. Since computation of derivatives is very much simpler than the laborious direct evaluation of areas (this was certainly true in the days before computers!) this formulation provided a highly convenient quick method for working out definite integrals of many simple functions – all those which were known to be derivatives of other functions – and thus it quickly converted the calculus from a difficult and esoteric discipline into a useful easy-to-handle mathematical tool. But the method of the antiderivative did not provide a *systematic* way of carrying out quadratures; quite simple functions like e^{-x^2} cannot be written as derivatives of other simple functions. It was only about 200 years later that a return was made, by Riemann and others, to the *direct* definition of the integral $\int_a^b f(x)\,dx$, in the Greek manner, as the limit of a sum without any reference to differentiation. The fact that

$$\frac{d}{db} \int_a^b f(x)\,dx = f(b)$$

then becomes a basic theorem in analysis; it is still known as the 'fundamental theorem of the calculus'. The more powerful nineteenth-century concept of the integral as a 'sum' rather than an antiderivative thus includes the earlier approach as a special case, and in the present age of computers it should be the natural way of thinking about integrals in practical as well as in theoretical terms. Unfortunately the teaching of elementary calculus in our schools is, even now, all too often still dominated by the thinking of the seventeenth century.

We return to Fermat and early notions of differentiation. The determination of maxima and minima, like the problem of tangents, suggested methods equivalent to finding the derivative of a function. Take the classical geometrical result that, of all rectangles with a given perimeter, the square has the greatest area. The following purely geometrical proof is given by Euclid (Fig. 9.9): suppose $A + C$ is a rectangle and $B + C$ a square of the

Fig. 9.9

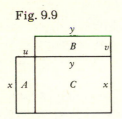

same perimeter. Since the perimeters are equal, $x + u + y = x + v + y$, so $u = v$. Since $B + C$ is a square, $y = v + x$, so $y > x$, hence the area of B (yv) is greater than the area of A $(xu = xv)$, and therefore the area of the square $B + C$ is greater than the area of the rectangle $A + C$. Fermat uses instead a (modern) analytical method: he notes that the problem amounts to finding the value of x for which the *area function* $J(x) = (a - x)x$ has its greatest value (Fig. 9.10). To do this Fermat exploited the idea, which

Fig. 9.10

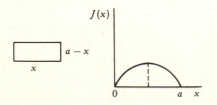

was not a new one, that near a maximum a function varies particularly slowly; thus $J(x)$ will change only very little when x changes by a small amount, so that we have a *stationary value* of the function. So, replacing x by $x - e$ where e is small, we form

$$J(x - e) = [a - (x - e)](x - e)$$

and put it equal to $J(x)$:

$$(a - x)x = (a - x + e)(x - e).$$

This equality should be more nearly true the closer e is to zero. Simplifying, we have

$$(a - x)x = (a - x)x - e(a - x) + ex - e^2$$

or

$$2ex - ea - e^2 = 0$$

or (dividing by e)

$$2x - a - e = 0.$$

Now setting $e = 0$ we have $x = a/2$ and hence also $a - x = a/2$; thus the rectangle with maximum area is a square. This procedure is of course entirely equivalent to our modern method of setting the first derivative $dJ(x)/dx$ equal to zero.

Even earlier, around 1610, the great astronomer Kepler had considered a similar problem, arising from the measurement of volumes of wine barrels. (He even published a book on the subject, the *Stereometria Doliorum*.) Kepler objected to his wine merchant's way of calculating the price of the wine; the measure of the price was the length l of a measuring rod which was inserted into the cylindrical barrel through a hole half-way up and made to rest against the top edge opposite (Fig. 9.11). Kepler realised that many different volumes could correspond to a given l (some very small!), and he worked out, for given l, the barrel height h_m for which the volume is a maximum. (*Exercise*: show that $h_m = 2l/\sqrt{3}$.)

The concept of differentiation arose also in a physical context in *kinematics*, the study of the speeds of moving bodies – a major preoccupation of seventeenth-century science. Galileo, Kepler's great contemporary, found in his experiments on bodies moving down inclined planes under gravity that the distance travelled varies as the square of the time, $x = at^2$. In trying to link this result with the behaviour of the speed, $v = dx/dt$, Galileo did not in fact proceed by differentiation; instead he tried various laws for the speed (first v proportional to x, later v proportional to t) and attempted to recover the distance/time law, by rather obscure reasoning, from the graph of speed against time (thus he was really trying to integrate the velocity function). A similar approach was used by Descartes (about 1618). Did Galileo and Descartes realise that the distances x were proportional to the *areas* under the velocity/time graph? We cannot be sure; a clear enunciation of the relation $x = \int_0^t v \, dt$ was given only about 40 years later, by Newton's teacher Isaac Barrow (see below). Certainly during this period it came to be more and more clearly accepted that functions and curves defined kinematically – i.e. the positions, orbits and speeds of particles as functions of the time – were mathematically indistinguishable from functions and curves defined in more general fashion, and that the time t was just a mathematical parameter equivalent to any other variable. Thus the approach to the calculus via mechanics became gradually fused with the approach via geometry.

We mention briefly some other major contributors to the early calculus before Newton and Leibniz:

Blaise Pascal (1623–62) was a highly original genius who contributed to many branches of mathematics. We have already

Fig. 9.11

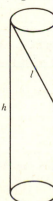
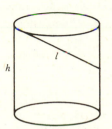

mentioned that he was a founder of the mathematical theory of probability and that he designed and built the earliest calculating machine. He made an intensive study of the properties of the cycloid, and he knew how to integrate simple functions such as x^n and $\sin x$. The writings of Pascal greatly influenced the young Leibniz.

Christiaan Huygens (1629–95), a wealthy Dutchman living in Paris, was distinguished as mathematician, physicist and astronomer. He is the creator of the wave theory of light. His book on pendulum clocks, the *Horologium Oscillatorum* (published in 1673), and his letters are rich in new results on properties of curves. Many of these were found only after Leibniz had already published his general method for tackling such problems. Huygens found it difficult to accept Leibniz's calculus, probably because of Huygens's insistence (unusual among mathematicians of the seventeenth century) on the high standards of rigour demanded by the Greek geometrical tradition of Archimedes.

John Wallis (1616–1703) was Savilian Professor of Geometry in the University of Oxford and helped to organise the Royal Society (founded 1660).[†] Wallis's *Arithmetica Infinitorum* (published in 1655) represents, together with Huygens's *Horologium*, the most highly developed form of infinitesimal calculus in the period before Newton and Leibniz. As the title shows Wallis wanted to demonstrate the power of the new 'arithmetic' (i.e. algebra), not the ancient geometry, and his bold methods in the treatment of infinite series and products gave him many new results. A typical formula obtained by Wallis is his famous infinite product for π. He derived this by writing $\frac{1}{4}\pi$ as the area of a quadrant of the unit circle $x^2 + y^2 = 1$, i.e. (in

---

[†] The Royal Society of London was one of the scientific academies which developed in the sixteenth and seventeenth centuries from discussion groups of learned men. The academies expressed the new spirit of free investigation, typifying 'this age drunk with the fullness of new knowledge, busy with the uprooting of superannuated superstitions . . .' (Ornstein). In this they contrasted with the universities which, founded in the Middle Ages, tended to maintain the medieval attitude of presenting knowledge in fixed forms. The earliest academies were Italian: Naples 1560, Rome (the 'Accademia dei Lincei') 1603.

modern notation) as the integral $\int_0^1 (1-x^2)^{1/2}\,\mathrm{d}x$. The square root in the integrand caused trouble, as there was no binomial theorem for general non-integral powers. Wallis therefore evaluated $\int_0^1 (1-x^2)^n\,\mathrm{d}x$ for $n = 0, 1, 2, \ldots$, and guessed the answer for $n = \frac{1}{2}$ by a complicated interpolation process.† This led him finally to his result

$$\frac{\pi}{2} = \frac{2 \times 2 \times 4 \times 4 \times 6 \times 6 \times 8 \times \cdots}{1 \times 3 \times 3 \times 5 \times 5 \times 7 \times 7 \times \cdots}.$$

James Gregory (1638–75) was a Scot who studied in Italy and who was in possession of large elements of the calculus by 1668. Unfortunately he used geometric rather than analytic methods which made his work difficult to follow. The infinite series

$$\tan^{-1} x = x - \frac{x^3}{3} + \frac{x^5}{5} - \frac{x^7}{7} + \cdots$$

is known as 'Gregory's series'; it was obtained by writing $\tan^{-1}x$ as an integral,

$$\int_0^x \frac{\mathrm{d}y}{1+y^2} = \int_0^x (1 - y^2 + y^4 - y^6 + \cdots)\,\mathrm{d}y,$$

and integrating the series term by term. Gregory knew the fundamental expansion known as the *Taylor series*:

$$f(a + x) = f(a) + xf'(a) + \frac{x^2}{2!} f''(a) + \cdots,$$

long before Brook Taylor published it in 1715.‡

Isaac Barrow (1630–77) was Newton's predecessor in the Lucasian chair at Cambridge (he resigned in 1669 and proposed Newton as his successor). Barrow's *Geometrical Lectures*

---

† In fact Wallis considered, more generally, the integral of $(1 - x^{2/\mu})^n$ for various integer values of both μ and n. For details see Edwards (1979).

‡ Not to mention Maclaurin: Colin Maclaurin's series

$$f(x) = f(0) + xf'(0) + \frac{x^2}{2!}f''(0) + \cdots,$$

which is only a special case of the Taylor series, was published in 1742. The naming of theorems is often a matter of historical accident rather than historical justice! The history of the Taylor series is actually very complicated; historians of mathematics claim that a special case was already known in India before 1550.

appeared in 1670 and were much concerned with tangent problems and quadratures. We have mentioned that Barrow saw clearly that the distance travelled in rectilinear motion is proportional to the area under the velocity/time graph. From his kinematical studies he must have been aware of the relation between the derivative regarded as the slope of a tangent and the integral regarded as an area. One of Barrow's results which involves this relation is the following: suppose two curves $y = f(x)$, $Y = F(x)$ are such that the ordinates Y are proportional to the areas $\int_c^x y \, dx'$, i.e. $aF(x) = \int_c^x f(x') \, dx'$, then the tangent at (x, Y) to $Y = F(x)$ cuts the x-axis at the point $x - T$, where T is given by $y/Y = a/T$. To obtain this theorem Barrow argued with small quantities in a manner similar to Fermat's; we would now obtain it at once by observing that the slope of the tangent is y, the derivative of Y. Barrow gives many results of similar type, but the *general* significance of the relation between integration and differentiation is not made clear by him, and his insistence on old-fashioned geometric language made his *Geometrical Lectures* difficult for others to follow.

The 1650s and 1660s were thus a period of rapid advance in the handling of a variety of infinite methods. Infinite series, infinite products and continued fractions were all in fashion. The publication of *Mercator's series* (1668)[†]:

$$\log(1+x) = \int_0^x \frac{dy}{1+y} = \int_0^x (1 - y + y^2 - y^3 + \cdots) \, dy$$

$$= x - \frac{x^2}{2} + \frac{x^3}{3} - \frac{x^4}{4} + \cdots$$

made a great impression; the realisation that a non-algebraic function like a logarithm could be represented in this way as a simple power series drew the attention of mathematicians to the use of infinite series as a general method for studying functions of all kinds. (Newton himself contributed the first enunciation of the binomial theorem for general non-integral powers.) Many

† Nicolaus Mercator (1620–87) must not be confused with the geographer Gerard Mercator (1512–94) who introduced 'Mercator's projection' in map-making.

integrations had been done by this time, differentiation was known and the relation between the two had been recognised; various techniques were known for relating one integral to another (thus Barrow had the rule for integration by 'change of variables'); even some problems equivalent to the solution of simple differential equations had been solved. The time was ripe for the creation of a general algorithm which could handle, in a single unified notation, all the fundamental operations of the infinitesimal calculus. This is what Newton and Leibniz achieved.

Newton and Leibniz were among the greatest minds of all time, and their creation of the calculus formed only a part of their many profound contributions to knowledge. There has been much (too much!) discussion about who should have the credit for being the first with the calculus; it is now generally agreed that the discoveries were made independently. In fact Newton's work (in 1665–6) antedated that of Leibniz (in 1673–6), but Leibniz was the first to publish, and his work exerted a much greater *immediate* influence on the mathematical thinking of the time.

Isaac Newton was born in 1642 (the year in which Galileo died) in the village of Woolsthorpe in Lincolnshire. He studied in Cambridge under Isaac Barrow and succeeded his teacher as Lucasian professor in 1669. Newton's most famous work is the *Principia* (the full title is *Philosophiae Naturalis Principia Mathematica*, published in 1687). In this unique work Newton establishes the laws of mechanics on an axiomatic basis and derives the laws of planetary motion from the universal inverse-square law of gravitation. The approach laid down in the *Principia* has become the model for all subsequent developments in physical science.

Newton's chief discoveries, including his calculus – which he called the 'method of fluxions' – were made during 1665 and 1666 when he had retired to his native village to escape from the plague in Cambridge. Rather surprisingly, the demonstrations in the *Principia* do not use the calculus but are based on 'old-fashioned' geometrical methods, presumably because

Newton considered these to be more logically satisfying. Throughout his life Newton was reluctant to publish his discoveries, and this makes the extent of his influence on his contemporaries difficult to assess. The *Method of Fluxions* was actually not published until 1736, after Newton's death.

Newton thought in mechanical terms: curves were generated by the continuous *motion* of a point. The coordinates (x, y) of a point on a plane curve are called 'fluents' (flowing quantities), their rates of change are 'fluxions', denoted by $(\dot{x}, \dot{y})$ (these are our modern derivatives dx/dt, dy/dt, with t as the time parameter; thus the fluxions are velocity components). The fluxion of $\dot{x}$ is $\ddot{x}$, and so on. The 'moment' of the fluxion $\dot{x}$ is the infinitesimal quantity $\dot{x}o$, where o is an 'infinitely small quantity'. Newton argues typically as follows: given, say, a plane curve $x^2 - axy - y^2 = 0$, substitute $x + \dot{x}o$ for x, $y + \dot{y}o$ for y, and so obtain

$$x^2 + 2x\dot{x}o + \dot{x}^2o^2 - axy - ax\dot{y}o - ay\dot{x}o$$
$$- a\dot{x}o\dot{y}o - y^2 - 2y\dot{y}o - \dot{y}^2o^2 = 0.$$

Now $x^2 - axy - y^2 = 0$; expunge this and divide the rest by o; there remains

$$2x\dot{x} + \dot{x}^2o - ax\dot{y} - ay\dot{x} - a\dot{x}\dot{y}o - 2y\dot{y} - \dot{y}^2o = 0.$$

'But whereas zero is supposed to be infinitely little, that it may represent the moments of quantities, the terms that are multiplied by it will be nothing in respect to the rest; I therefore reject them, and there remains

$$2x\dot{x} - ax\dot{y} - ay\dot{x} - 2y\dot{y} = 0.'$$

(This is the calculus result, x and y being differentiable functions of t.)

What kind of quantities are the os? Are they zeros? Or 'infinitesimals' (whatever that means)? Or finite numbers? Newton tried to explain their nature by a primitive notion of a limit, but his argument (from the *Principia*) is hardly clear:

'Those ultimate ratios with which quantities vanish are not truly the ratios of ultimate quantities, but limits toward which the ratios of quantities, decreasing without limit, do always converge, and to which they approach nearer than by any given difference, but never go beyond, nor in effect attain to, until the quantities have diminished in infinitum.'

Newton in practice knew well enough what he wanted to do with his calculus and was able to use it in the absence of clear definitions; but lesser mortals were more easily confused.

Gottfried Wilhelm Leibniz was born in Leipzig in 1646 and spent most of his life at the court of Hanover, in the service of the dukes (one of whom became King George I of England). He was the great universal genius of the seventeenth century, with interests in history, theology, linguistics, biology, geology, diplomacy and mathematics; he invented a computing machine, conceived the idea of a steam engine, studied Sanskrit and tried to promote the unity of Germany. Leibniz met Huygens in Paris in 1672 and took lessons in mathematics from him. He developed his calculus between 1673 and 1676 and published it between 1684 and 1686. Leibniz was as much philosopher as scientist and wanted to find a 'universal language' to describe all change (motion included), and beyond this a universal method for acquiring knowledge and for making new inventions. A modern reader may feel sceptical about such sweeping ambitions, but Leibniz's attitude led to important results: he laid the foundations for symbolic logic, and he was aware, more than any of his contemporaries, of the importance of a well-devised and easily handled mathematical *notation*. The symbolism he introduced for the calculus made it easy for others to understand and handle his methods, and his notation is in fact the one that has survived to the present day. Leibniz's first paper on the calculus contained our symbols dx, dy, the product rule $d(uv) = u\,dv + v\,du$ and the condition $dy = 0$ for extreme values; later he introduced the integration sign $\int$ (which is a long S, denoting 'summa'); and the names 'differential calculus' and 'integral calculus' are also due to him.†

The publication of Leibniz's calculus initiated a period of extremely fertile productivity. Two very able disciples who

---

† Leibniz's first suggestion for the integral calculus was 'calculus summatorius', later replaced by 'calculus integralis'. Modern analysis has re-adopted Leibniz's earlier suggestion.

eagerly took up Leibniz's methods were the brothers Jakob and Johann Bernoulli. Working together – and often in bitter rivalry – these two had by 1700 discovered most of the material in the present-day A-level calculus syllabus, together with portions of more advanced subjects such as the solution of many ordinary differential equations and results in the calculus of variations. In 1696 the first textbook on the calculus (the *Analyse des infiniment petits*) was published by the Marquis de l'Hospital; this was essentially a text written a few years earlier by the Marquis's teacher Johann Bernoulli. The Marquis is remembered today by 'L'Hospital's rule', given in this book, for finding $\lim_{x \to 0} f(x)/g(x)$ when f and g both tend to zero.

Leibniz's calculus was just as vague as Newton's about the logical foundations; he, also, could not say clearly whether the quantities dx, dy were to be regarded as finite or zero or as something in between.

Two further comments should be made.

(1) The lack of precise foundations for the new calculus provoked criticism. The best known came from Bishop Berkeley, the Irish idealist philosopher, who resented Newtonian science because it supported materialism, and who attacked the theory of fluxions in his *Analyst* (published in 1734). Berkeley was a lively debater: he derided infinitesimals as 'ghosts of departed quantities' and called Newton's arguments, in which o is sometimes taken as non-zero and sometimes as zero, a 'manifest sophism'. According to Berkeley, 'he who can digest a second or third fluxion, a second or third difference, need not, methinks, be squeamish about any point in divinity'. Although these and similar criticisms had justification, they were purely destructive and did not supply any better basis for the calculus. They did however underline the need for further work to be done on the foundations of the subject.

(2) There were prolonged disputes as to whether Newton or Leibniz deserved the priority for discovering the calculus, disputes which the followers of the two great men pursued with zeal and venom, and which were made disagreeable by accusations of plagiarism.

These criticisms and quarrels contributed to the generally feeble state of British mathematics in the eighteenth century, at a time when enormous mathematical advances were made in Europe with the aid of the calculus. No doubt other factors were also to blame for Britain's weakness, such as an over-emphasis on old-fashioned geometric concepts and a general lack of public regard for the importance of mathematical studies. An over-pious adherence to all the details of Newton's fluxional methods led to revolt in 1812 when the 'Analytical Society' was formed by young mathematicians in Cambridge to promote Leibniz's notation and to reform British mathematics generally. But it was a long time before the writings of the great continental analysts came to be fully understood in England (see Hardy 1949).

10

The function concept

The new calculus was developed with vigour in the eighteenth century, mainly on the Continent of Europe. The foremost name was Euler who was Swiss but spent most of his life in St Petersburg and Berlin; his mathematical output was immense. The calculus provided a powerful tool for developing Newton's mechanics and applying it to all sorts of physical systems. For this purpose people learned to handle *differential equations* with skill and a highly developed technique. Because of the success of the calculus in leading to physical theories which made sense, mathematicians did not worry too much about the fact that the rigorous foundation was lacking. It was a period of happy and bold experimentation in mathematics. But there was no lack of argument about the meaning and validity of the new developments.

We shall pick out for brief discussion a topic important in the creation of present-day analysis: early controversies surrounding the concept of a *function*. For Euler and his contemporaries functions were entities like e^x, $\log(1+x)$, $\sqrt{(1+x)}$; they could all be represented by a *formula*. Thus when faced with (for example) the expressions

$$y = x \, (x \leqslant 0), \qquad y = x^2 \, (x > 0),$$

Euler would say that we have here *two* functions because we have two formulae. Alternatively, looking at the graph of y against x (Fig. 10.1), Euler might say that we have one function (since there is only one curve), but that it is 'discontinuous'. Nowadays we have a much more general view of the concept of a function. It is just a rule which assigns to each element of a set A a unique element in a set B. Thus for us the above example is a single function (which, according to modern definitions, is continuous and has a discontinuous derivative at $x = 0$); the fact that two different formulae are needed is an aspect of minor significance. (We learn this in analysis but sometimes forget about it in applied mathematics.) Note also that Euler's functions usually had the property, much used in the eighteenth-century calculus, that they could be expanded as infinite series in powers of x. For example:

$$e^x = 1 + x + \frac{x^2}{2!} + \cdots,$$

$$\log(1+x) = x - \tfrac{1}{2}x^2 + \tfrac{1}{3}x^3 - \cdots,$$

$$\sin x = x - \frac{x^3}{3!} + \frac{x^5}{5!} - \cdots.$$

Euler and his contemporaries were ingenious in manipulating infinite series. Usually they did not worry about convergence, and some of the results do not make any sense to us. Thus Euler somewhere argues as follows: since

$$\frac{1}{1-x} = 1 + x + x^2 + \cdots, \quad \text{one has} \quad \frac{x}{1-x} = x + x^2 + x^3 + \cdots$$

and

$$\frac{x}{x-1} = \frac{1}{1-(1/x)} = 1 + \frac{1}{x} + \frac{1}{x^2} + \cdots = -\frac{x}{1-x}.$$

So, adding, one obtains the expansion

$$\cdots + \frac{1}{x^2} + \frac{1}{x} + 1 + x + x^2 + \cdots = 0.$$

We would call this a *formal* result which is meaningless: since the first series converges only for $|x| < 1$ and the second only for $|x| > 1$, the two series cannot be added.

Fig. 10.1

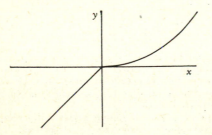

For us the *sum* of an infinite series is a *limit* and exists only when the limit exists.† Although eighteenth-century concepts of the sum of an infinite series were not precise, it should not be thought that questions of convergence were completely ignored. Mathematicians were well aware of the convergent or divergent nature of particular series: they knew for example that the 'harmonic series' $\sum_1^\infty 1/n$ diverges while $\sum_1^\infty 1/n^2$ converges.‡ They were particularly interested in the use of series to obtain good approximations in numerical calculations, and some important methods for evaluating sums of the form $\sum_{k=1}^n f(k)$ when n is large were first given by Euler. Although they often could not prove their results rigorously, the mathematicians of the period were guided by an excellent intuitive feel for the correct approach.

A particular problem solved by Euler was that of finding the sum of the convergent series

$$\sum_1^\infty \frac{1}{n^2} = \frac{1}{1^2} + \frac{1}{2^2} + \frac{1}{3^2} + \cdots,$$

a problem that had defeated Leibniz and Jakob Bernoulli. Euler not only obtained accurate numerical *estimates* for this sum, but he also evaluated the sum *exactly* by the following bold approach. He first considered the algebraic equation of degree n

$$(x - x_1)(x - x_2)(x - x_3) \cdots (x - x_n) = 0$$

with roots $x_1, x_2, \ldots, x_n$, where we are given that the constant term is equal to 1, so that

$$(-1)^n x_1 x_2 \cdots x_n = 1. \tag{1}$$

The coefficient of x in the equation is

$$(-1)^{n-1}(x_2 x_3 \cdots x_n + x_1 x_3 \cdots x_n + x_1 x_2 \cdots x_{n-1}),$$

and this, dividing by (1), is seen to be equal to

$$-\left(\frac{1}{x_1} + \frac{1}{x_2} + \cdots + \frac{1}{x_n} \right).$$

---

† We recall that a series $u_1 + u_2 + \cdots + u_n + \cdots$ is convergent with sum s when the sum of n terms $(u_1 + u_2 + \cdots + u_n)$ tends to the limit s as $n \to \infty$. More general definitions of the sum of an infinite series can be given which allow non-convergent series to be summed in a generalised sense.

‡ It seems that the proof that the harmonic series is divergent goes back to the medieval scholar Nicole Oresme (1323?–82).

Euler's idea was to apply this result to the equation $\sin x = 0$. Using the series for $\sin x$ he regarded this as an algebraic equation of *infinite* degree:

$$x - \frac{x^3}{3!} + \frac{x^5}{5!} - \cdots = 0,$$

or (dividing by x and putting $x^2 = w$)

$$1 - \frac{w}{3!} + \frac{w^2}{5!} - \cdots = 0. \tag{2}$$

The non-zero roots of $\sin x = 0$ are known to be the infinite set $x = \pm\pi,\ \pm 2\pi,\ \pm 3\pi, \ldots$, so we know that the roots of (2) are π^2, $(2\pi)^2$, $(3\pi)^2, \ldots$. Also the constant term is 1, and the coefficient of w is $-1/6$. Hence we have for the sum of the reciprocals of the roots

$$-\left(\frac{1}{\pi^2} + \frac{1}{(2\pi)^2} + \frac{1}{(3\pi)^2} + \cdots\right) = -\frac{1}{6},$$

or

$$\frac{1}{1^2} + \frac{1}{2^2} + \frac{1}{3^2} + \cdots = \frac{\pi^2}{6}.$$

This elegant formula is the correct result. Euler's derivation would not be accepted nowadays without a careful justification of the transition from finite n to infinite n, but it was a brilliant achievement in its time. Alternative derivations confirming the result were also given by Euler.

A particular physical problem much discussed in the eighteenth century contributed greatly to the formulation of the modern concept of a function. The problem is that of calculating the shape of a *vibrating string* (for example a violin string), fixed at its ends and undergoing small transverse oscillations. The mathematical task is to find the function $y(x, t)$ which specifies the transverse displacement of the string at a distance x along the string at time t (Fig. 10.2). y is thus a function of the

Fig. 10.2

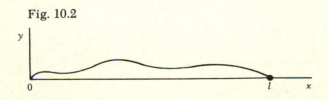

two variables x and t. If the ends of the string are at $x = 0$ and $x = l$, then, since these ends are fixed, y must satisfy the *boundary conditions*

$$y(0, t) = y(l, t) = 0 \quad \text{for all } t.$$

If the *initial* displacement of the string (at time $t = 0$) is given, by specifying some function $f(x)$, then y also satisfies the *initial condition* $y(x, 0) = f(x)$. Around 1747 d'Alembert and Euler solved the partial differential equation (the *wave equation*) satisfied by $y(x, t)$, and they showed that, with the specified initial and boundary conditions, the problem has a well-defined unique solution which can be expressed in the following elegant form. The wave equation, together with the boundary condition at $x = 0$, can be satisfied by writing y in the form

$$y(x, t) = F(ct + x) - F(ct - x), \tag{3}$$

where $F(u)$ is any 'function'. Physically the two expressions on the right-hand side represent *waves* travelling along the string to the left and to the right, where the constant c is the *wave velocity*. The function F is determined by the initial displacement of the string. We must have $f(x) = y(x, 0) = F(x) - F(-x)$; suppose we define $F(u)$ to be an *odd* function of u (so that $F(u) = -F(-u)$), then this gives us $F(u) = \frac{1}{2}f(u)$. For example, suppose the initial displacement of the string is a pure sine curve, with $f(x) = 2\sin(\pi x/l)$, then the solution (3) is

$$\sin\frac{\pi}{l}(ct + x) - \sin\frac{\pi}{l}(ct - x) = 2\sin\frac{\pi x}{l}\cos\frac{\pi ct}{l}. \tag{4}$$

Fig. 10.3

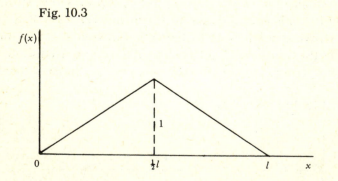

(Note that this expression also satisfies the boundary condition at $x = l$.) The second form of the solution shows that we have here a vibration with time dependence of the form $\cos 2\pi\nu t$, i.e. a *pure tone* with a single frequency $\nu = c/2l$. (This important special solution had been given earlier by Brook Taylor, around 1715.) To satisfy the end condition at $x = l$ in general we must require that $F(ct + l) = F(ct - l)$. This condition can be written $F(u) = F(u + 2l)$ for all u; thus $F(u)$ must be a *periodic function* of u with period $2l$. (This is certainly true of the function $\sin(\pi u/l)$, since $\sin(\pi u/l) = \sin(\pi/l)(u + 2l)$.)

Euler realised that the 'pure tone' solution is only a special case: for example, if we pull the string aside at its mid-point by a unit amount and let go (the case of a 'plucked string'), we have the initial shape illustrated in Fig. 10.3 and described mathematically by

$$
\left.
\begin{aligned}
f(x) &= 2x/l && (0 \leqslant x \leqslant \tfrac{1}{2}l), \\
&= 2(1 - x/l) && (\tfrac{1}{2}l \leqslant x \leqslant l).
\end{aligned}
\right\}
\tag{5}
$$

To what kind of function $F(u)$ does this initial condition lead, in the general solution (3)? Remember that $F(u)$ must be periodic with period $2l$, and that $f(u) = 2F(u)$ if $F(u)$ is an odd function of u. Thus we take $F(x) = \tfrac{1}{2}f(x)$ in the interval $0 \leqslant x \leqslant l$, and we *continue* it for other values of x as an odd function of x which has the required periodicity. We obtain the function illustrated in Fig. 10.4. With this choice of $F(x)$ Euler's solution $F(ct + x) - F(ct - x)$ satisfies all the conditions of the problem.

Fig. 10.4

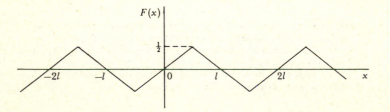

Daniel Bernoulli had quite a different way of tackling the same problem.† He used a more physical argument. It had been known since about 1700, from experimental work on vibrating strings, that *any* vibration could be regarded as made up of a superposition of a *fundamental tone* and *overtones* vibrating at frequencies which are multiples of the basic frequency. (The Pythagoreans had already, 2000 years earlier, connected the properties of whole numbers with music; they knew that when the lengths of strings are in the ratios of simple whole numbers the tones emitted are harmonious.) Mathematically the fundamental tone will be of the form of equation (4), and we must add to this all the overtones. Thus a general displacement $y(x, t)$ is on this view to be written as an infinite linear superposition of different modes of vibration as follows (the Ks being suitable amplitude coefficients):

$$y(x, t) = K_1 \sin \frac{\pi x}{l} \cos \frac{\pi c t}{l} + K_2 \sin \frac{2\pi x}{l} \cos \frac{2\pi c t}{l}$$

$$+ K_3 \sin \frac{3\pi x}{l} \cos \frac{3\pi c t}{l} + \cdots \qquad (6)$$

All the terms in this series are zero at $x = 0$ and $x = l$ as required – but what about the initial shape $f(x)$? Put $t = 0$ in (6), then the cosines are all equal to 1, and so we must have

$$f(x) = K_1 \sin \frac{\pi x}{l} + K_2 \sin \frac{2\pi x}{l} + K_3 \sin \frac{3\pi x}{l} + \cdots \qquad (7)$$

Does this make sense? More specifically, can a series of smooth sine functions possibly represent a function with a 'corner', such as the function (5) representing the initial displacement of a plucked string? The eighteenth-century mathematicians had a new and unfamiliar problem here, since the function (5) is not a simple function in Euler's sense, and the series (7) is not the

---

† The Bernoulli family holds the record in history for the number of distinguished mathematicians it produced. Daniel Bernoulli (1700–82), the son of Johann, was what we would call an 'applied mathematician'; he laid the foundations of the science of hydrodynamics (the application of Newton's mechanics to the flow of fluids) and of the kinetic theory of gases.

usual Taylor series expansion in powers of x. Euler's view was that Bernoulli's series (7) *cannot* represent the initial shape of the string in general, since a series of sines cannot 'add up' to a function with corners. His argument seemed convincing. After all, the series (7) is a single *formula* (even though it has an infinity of terms), and (5) is not; also each sine function is smooth, with no corners (we say that it has a continuous derivative) – the higher terms in (7) just oscillate faster (Fig. 10.5) – how can a series of such smooth oscillations possibly represent a function such as that in Fig. 10.4? Bernoulli was nevertheless firmly convinced that he had written down a general solution, and there was a lengthy controversy over just what kind of function can be represented by *trigonometric series* of the type (7). The argument went on throughout the eighteenth century, and several mathematicians (in particular Lagrange) came close to the answer, but the first man to give a clear resolution of the problem was Joseph Fourier, who lived from 1768 to 1830. He fully understood the situation but had no rigorous proofs for his results, and his ideas were regarded as so revolutionary that his first paper on the subject was rejected (in fact by Lagrange, in 1807). But Fourier was undaunted and finally published in 1822 his famous work *Théorie analytique de la chaleur* (where essentially the same problem is discussed in connection with heat flow). What was Fourier's insight; was Euler right or Bernoulli? Fourier realised that *both Euler's and Bernoulli's* solutions were correct, and that (surprising, but true) a function with 'corners' can indeed be represented by an infinite trigonometric series as in (7), now called a *Fourier*

Fig. 10.5

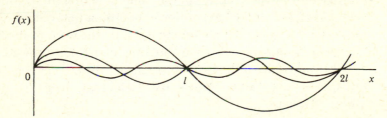

series.† Moreover, to represent $f(x)$ in (5) as the series (7), we must calculate the coefficients in accordance with the formula

$$K_n = \frac{2}{l} \int_0^l f(x) \sin \frac{n\pi x}{l} \, dx. \tag{8}$$

This integral is easily evaluated for our particular $f(x)$ to give the result

$$K_n = 0 \quad \text{when } n \text{ is even,}$$

and

$$K_{2m+1} = (-1)^m \frac{8}{(2m+1)^2 \pi^2} \quad \text{for } m = 0, 1, 2, \ldots$$

With these coefficients the series (7) represents the function (5) in the interval $(0, l)$. What function does it represent outside the interval? It is clearly an odd function of x (since the sines are odd functions), and it has the period $2l$ (since each term has this period); so we have just the function sketched in Fig. 10.4, required by Euler's solution!

If instead we had continued the function (5) as an *even* function of x, we would have obtained the function sketched in Fig. 10.6. This can also be represented as a Fourier series but a

Fig. 10.6

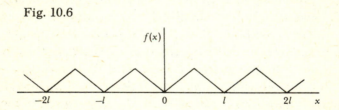

different one which uses cosines. Thus we have two different Fourier series which represent the same 'function' in the interval $0 \leqslant x \leqslant l$, but which have different continuations outside this interval. This example focuses attention on the need to specify the set of values on which a function is defined (nowadays called the *domain* of the function). Clearly the old concept of a function as something represented by a single formula is no longer adequate.

† In general a Fourier series will contain both sine and cosine terms.

Fourier went further, and claimed that one can represent not only functions with 'corners' (i.e. continuous functions with discontinuous derivatives), but also periodic functions which are themselves discontinuous, by trigonometric series. Take for example the function defined in $0 \leqslant x \leqslant l$ by

$$y = 1 \quad (0 \leqslant x \leqslant \tfrac{1}{2}l), \quad y = 0 \quad (\tfrac{1}{2}l < x \leqslant l),$$

shown in Fig. 10.7. Application of the formula (8) leads to the Fourier sine series

$$\frac{2}{\pi} \left(\sin \frac{\pi x}{l} + \sin \frac{2\pi x}{l} + \frac{1}{3} \sin \frac{3\pi x}{l} + \frac{1}{5} \sin \frac{5\pi x}{l} + \cdots \right), \qquad (9)$$

which represents the odd periodic function shown in Fig. 10.8.†
The claim that such a discontinuous function can be expanded as a Fourier series caused more controversy, involving a new and subtle point. Cauchy, the leading French mathematician of the early nineteenth century – one of the founders of rigorous analysis – claimed to have *proved* that any sum of a series of

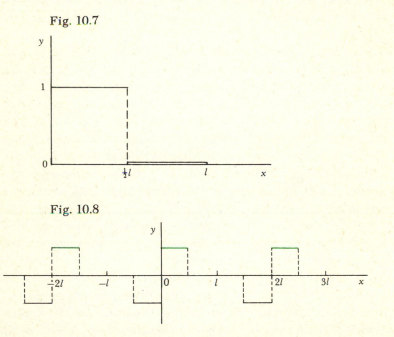

Fig. 10.7

Fig. 10.8

† The coefficient of the nth term is $(2/\pi n) (1 - \cos \tfrac{1}{2} n \pi)$.

continuous functions must itself always be continuous. If this were true, a Fourier series could not possibly represent a discontinuous step-function of the type shown in Fig. 10.8. Cauchy was wrong on this point, and Fourier's boldness was again justified; it is indeed possible for an infinite sum of continuous functions to have discontinuities (a remarkable result)! The need to settle this question led to the first careful study of the convergence properties of series of *functions of x*. This is a more difficult question than that of the convergence of series of numbers, and it was not fully cleared up until some time later. Several mathematicians contributed; the full theory was given around 1850 by Weierstrass who was the foremost architect of rigour in analysis. We have to say that, when we approach a point of discontinuity of $f(x)$, the convergence of the Fourier series is no longer *uniform* in x; which means (very roughly speaking) that the series converges more and more slowly the nearer x gets to the point of discontinuity. Cauchy's result does not apply to all convergent series, but only to those which are *uniformly convergent*, and thus the apparent contradiction is resolved.

The first satisfactory mathematical proofs of Fourier's main results were given by Dirichlet, around 1830. He showed for example that the Fourier series representing a discontinuous function converges to the value $\frac{1}{2}[f(x-0)+f(x+0)]$ at a point of discontinuity, i.e. to the average of the values on the left and the right of the discontinuity. To illustrate this, consider the Fourier sine series for the step-function of Fig. 10.8 at the origin $x = 0$. When $x = 0$ all the sines are zero, and the series adds up, not to $y = 1$, but to $\frac{1}{2}[1+(-1)] = 0$, in accordance with Dirichlet's theorem.

Fourier's work with its startling results helped to give impetus to the renewed emphasis in the nineteenth century on questions of rigour in mathematical proof and on the clarification of basic mathematical concepts. The central part of this story is the creation of analysis – the rigorous form of the infinitesimal calculus – to which many mathematicians contributed. It is beyond our scope to describe this, but we shall conclude the chapter by noting a few points of special interest.

Dirichlet (1805–59), who made many fundamental contributions to both pure and applied mathematics, gave as long ago as 1837 a general definition of a function which was close to our modern one. He suggested, as an example of a 'badly behaved' function $f(x)$, the function which has the value 1 whenever x is rational and 0 whenever x is irrational. This is a perfectly well-defined function but it is *everywhere* discontinuous. There can be no question of 'drawing the graph' of such a thing; it is far removed from Euler's concept of a function as something given by an explicit formula, with a smooth curve and a power series expansion.

The generalized view of the idea of a function made it essential to re-examine the concept of the *definite integral*. The old intuitive idea, deriving from the Greeks, of $\int_a^b f(x)\,dx$ as the 'area under the curve' is applicable without any difficulty to functions with 'corners' and discontinuous step-functions of the type we discussed in connection with Fourier series. We can always break up the region of integration into pieces if necessary, and evaluate the integral as the sum of separate areas, as indicated in Fig. 10.9. The rigorous formulation of the area concept is due to Riemann who made precise earlier work by Cauchy. It restored the definite integral as a primary concept of the calculus, no longer defined in terms of a mere antiderivative (compare our discussion on p. 114); as Fig. 10.9 indicates, the 'area' definition holds for functions which need not possess derivatives everywhere. In 1854 Riemann gave a strict analytical definition of a definite integral based on the old idea of breaking up the area into a large number N of small rectangular strips, taking the total area of these (a sum of N terms), and then

Fig. 10.9

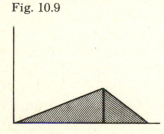
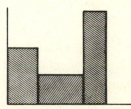

studying the limiting behaviour of such a sum as N tends to infinity and the widths of all the strips tend to zero. This allowed Riemann to define the integral for a wide class of functions and to state necessary and sufficient conditions for a function to be integrable. Riemann's work put the definite integral of the Newton–Leibniz calculus on a rigorous footing. But his definition could not be used for all the generalised functions of the new kind: thus for the completely discontinuous function introduced by Dirichlet the 'area under the curve' has no meaning, and Riemann's sum limit does not exist. If we want to define $\int_a^b f(x)\, dx$ for such functions, we need a more general definition of an integral. Many modern generalisations of Riemann's integral have been given, applicable to much wider classes of functions. A particularly important generalisation is that given by Lebesgue (about 1900): it requires a general (set-theoretic) definition of the length of an interval (called its *measure*) which allows us to say (for example) what 'fraction' of the length of the interval (0, 1) of the real axis is occupied by rational numbers. This is a subtle matter (we obviously cannot give the answer by measuring such lengths with a ruler); we shall return to it in the next chapter. A new branch of analysis called *measure theory* studies questions of this type and has given deep general insights into the concepts of function and integral. Although Dirichlet's function may seem an artificial invention, highly discontinuous functions frequently occur in applications of mathematics (see below), and measure theory is by no means only an abstract *jeu d'esprit*. It is needed for example to give a satisfactory mathematical formulation for the basic concepts of the *theory of probability*.

What about the notion of the derivative, and the problem of clarifying the exact nature of the 'infinitesimal quantities' (Newton's o, Leibniz's dx and dy) which had remained obscure since the seventeenth century? Definitions of a continuous function and of the derivative of a function were given by Cauchy in a series of influential textbooks (1820–30). Cauchy's concepts were essentially those used nowadays, although his formulations were not yet as precise as we would require today. His basic notion was that of a limit (50 years earlier d'Alembert had already insisted on this as the fundamental idea), and the derivative of $y = f(x)$ was defined as

$$f'(x) = \lim_{h \to 0} \frac{f(x+h) - f(x)}{h}.$$

This is just our modern expression, and Cauchy's attempt to define what is meant by a limit was also similar in spirit to the modern definition: 'When the successive values attributed to a variable approach indefinitely a fixed value so as to end by differing from it by as little as one wishes, this last is called the limit of all the others.' In this approach the mystery of the infinitesimals is entirely contained in the notion of the limit. The significant step forward was to regard an infinitesimal no longer as a *fixed* number, with all the difficulties of defining its ('infinitely small'?) magnitude, but as a *variable* number which can take values as small as one wishes. In this way it was possible to give a precise meaning to dy/dx, and there was no longer any need to worry about the nature of the quantities dy and dx as separate entities. In a modern calculus course one is taught very firmly that dy/dx is not a *ratio* but a *limit*, and that dy and dx have no separate meaning.† Has this solved the problem of the nature of the Newton–Leibniz infinitesimals, or has it merely side-stepped it? The reader is invited to form his own view. The formulation of the calculus in terms of limits is the one generally adopted and taught nowadays, but it has not satisfied all mathematicians, and there has always been a feeling that it should be possible to give a precise formulation of the calculus based directly on Newton and Leibniz's 'infinitesimally small' quantities as 'numbers' in their own right. This can indeed be done nowadays, and such a theory was worked out in detail about 20 years ago by A. Robinson who called the subject 'non-standard analysis'. It is based on an extension of the real

---

† Once $f'(x)$ has been determined, one can if one likes define differentials dy and dx as *finite* quantities which are such that their ratio equals $f'(x)$: thus we can write $dx = 1$, $dy = f'(x)$; but this of course adds nothing new. Physicists like to think in terms of finite small increments δx, δy whose ratio $\delta y/\delta x$ is *approximately* $f'(x)$; mathematicians generally prefer to avoid the use of finite differentials. Cauchy himself still tended to think of dy and dx as 'infinitely small' quantities; see Robinson (1966), ch. 10.

number system – the 'hyperreal numbers' – for which the Archimedean axiom (see Chapter 8) does not hold and which includes 'infinitely large' and 'infinitely small' numbers. An elementary 'non-standard' account of the calculus has been given by Keisler (1976).

Once satisfactory definitions had been given of continuity and differentiability, it began to be realised that a *continuous* function (whose curve has no breaks or jumps, so that it can be drawn 'without raising the pencil from the paper') is not necessarily also *differentiable* (i.e. it may not possess a gradient or tangent anywhere). It was certainly realised early on that a function such as that illustrated in Fig. 10.3, whose graph has a 'corner', will have no tangent at that point; but it was generally assumed that, apart from the occurrence of such isolated corners, a function which was continuous must also be differentiable. Various mathematicians in the nineteenth century realised that this was not necessarily so and gave examples of functions which were continuous everywhere but differentiable nowhere. It seems that the first example was given by Bolzano in 1834. (Bernhard Bolzano, 1781–1848, was a Czech scholar who found for himself many of the concepts of rigorous analysis well before others did so. As he worked in Prague, far from the chief centres of mathematics, his work was hardly noticed in his lifetime.) Later examples of continuous functions without tangents were given by Riemann and Weierstrass and attracted more attention. The peculiar behaviour of

Fig. 10.10

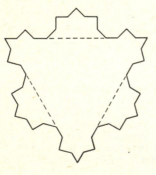

such functions cannot be fully illustrated by drawing a graph: roughly speaking, the 'graph' will *look* smooth, but close inspection will reveal that it is extremely 'crinkly', with a 'corner' or 'prickle' at every point. Such curves can be constructed by elementary methods; a simple example is the 'snowflake curve' suggested by von Koch in 1904. We start with an equilateral triangle and construct prickles by trisecting the sides and erecting further equilateral triangles on the middle portions (Fig. 10.10). This process of subdivision is to be continued indefinitely, so that ultimately there is a prickle at every point of the curve. Obviously many constructions along similar lines are possible.

Such creations may, again, strike the reader as mere mathematical curiosities, not likely to be of interest to the serious-minded mathematician or scientist. Many nineteenth-century mathematicians trained in the classical tradition felt this repugnance; thus Hermite wrote: 'I turn away in horror from this regrettable plague of continuous functions that do not have a derivative at even one point'; and there was an equal lack of interest on the part of physicists and applied mathematicians, used to dealing with projectiles and planets which travel smoothly along orbits according to the laws of Newtonian mechanics. But this attitude was quite misconceived: there are many examples in nature of highly irregular configurations for which continuous non-differentiable functions can provide a useful description. Imagine for example measuring a coastline, first along every headland and beach, then around every rock,

Fig. 10.11

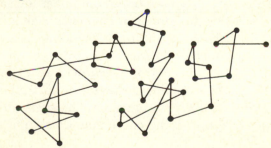

around every grain of sand on the shoreline, and so on. The irregular boundary of a cloud in the sky, and the turbulent dispersion of a blob of black ink in water, are two other examples, and the reader will be able to think of many more: it is reasonable to assert that irregular configurations are the rule and not the exception in nature. (Many examples are given by Mandelbrot (1977) who has coined the term 'fractals' to describe all such irregular shapes.) A characteristic feature of such configurations is that they grow more and more irregular the more closely we look at them: our image depends on the magnification employed and the method of measurement (when we represent the coastline on a map, the irregularity we see depends on the scale employed). The snowflake curve exhibits this property, but it is too regular in form to serve as a realistic model for naturally occurring irregularities. An important example of highly irregular paths in nature is provided by the *Brownian motion*: this refers to the tracks of small particles in a suspension which are suffering deflections caused by continual collisions with molecules which are themselves in irregular random motion (Fig. 10.11). Here the picture of the paths becomes more and more irregular the smaller the time interval at which successive positions are observed. We cannot in practice hope to have full information on the details of the path followed by any particle, and our theoretical description must make use of the concepts of the theory of probability. Thus the problem is treated mathematically as a 'random walk' in which we may assume (for example) that all paths between collisions are of equal length (which can be made infinitely small), but that all directions of motion after a collision are equally probable. The mathematical study of the paths traced out under such assumptions was initiated by the American mathematician N. Wiener (around 1920), and the theory of Brownian motion has become a major and active branch of the theory of probability.

Thus 'fractals' should not be regarded as pathological mathematical monsters, but they do have unusual mathematical properties. Consider for example the total *length* of Koch's snowflake curve: although the curve bounds a finite area, its length is multiplied by a factor $\frac{4}{3}$ at each of the subdivisions, so that after n steps the original length has increased by a factor $(\frac{4}{3})^n$, and this factor increases without limit as n tends to infinity; thus the length is clearly infinite! We are also forced to re-examine the intuitive idea of *dimension*. This question became acute when Peano in 1890 gave an example of a continuous non-differentiable curve which was so 'winding and twisting' that it passed through *every point* of a square area. Mathematicians had always regarded *curves* as one-dimensional configurations and *areas* (or *surfaces*) as two-dimensional (and the number of dimensions could be defined quite simply as the smallest number of parameters required to describe the figure). Peano's 'space-filling' curve seemed to remove the distinction between curves and surfaces; what dimension should be ascribed to it? The answer is: 'it depends how one defines dimension'; alternative definitions are possible and may give different answers in the case of irregular configurations. We briefly outline one possible approach, related to one suggested by the German mathematician Felix Hausdorff (in 1919). It applies to 'self-similar' figures, where we can reduce the scale of the figure in some ratio r (with $1/r$ equal to some integer), and obtain a reduced figure which is of the same form as the original one. Suppose that in this way we subdivide the original figure into N similar parts. In the case of a straight line of unit length we obviously obtain N segments each of length r, with $Nr = 1$; in the case of a unit square we obtain N squares each of area r^2, with $Nr^2 = 1$ (Fig. 10.12). A unit cube will give N cubes each of volume r^3, with $Nr^3 = 1$; in the general case of a D-dimensional figure we shall clearly expect a subdivision for which $Nr^D = 1$.

Fig. 10.12

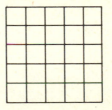

To solve this equation for D we take logarithms and find

$$\log N + D \log r = 0, \qquad \text{or } D = -\frac{\log N}{\log r} = \frac{\log N}{\log (1/r)}.$$

All this is simple enough, but note that, on this definition, D does not necessarily have to be an integer. An example of a shape for which D is non-integral is the snowflake curve, Fig. 10.10. By virtue of its construction this is a self-similar configuration, but a subdivision in the ratio $r = \frac{1}{3}$ now clearly gives us $N = 4$ similar parts! Thus in this case $D = \log 4/\log 3$, and so the dimension of the snowflake curve is a number which lies between 1 and 2. With other constructions other non-integral values of D can be obtained (see Mandelbrot 1977); while for Peano's space-filling curve it is found, as might be expected, that $D = 2$. But, as we have mentioned, other definitions of dimension can be given; and the general theory of the concept of dimension is a deep problem in topology which has not yet been fully resolved.

We note finally that space-filling curves also provide useful models of naturally occurring structures. Examples are 'branching networks' such as a river system, where we consider a main river together with all its tributaries, then the tributaries of the tributaries, and so on down to the very smallest side-streams; or a vascular system such as the complex intercommunicating set of vessels in the human lung. The mathematical model of such structures is a 'tree' like the one illustrated in Fig. 10.13, where we start with a 'trunk' which divides at one end into two branches, each of these branches dividing into two, and so on indefinitely. With such a network we can reach any point in the plane.† Suppose we regard the network as a model of a river system (all streams having zero width), and we follow the river bank continuously from the point A on the river's mouth until we eventually return to B on the opposite side: evidently we have traced out a single plane-filling Peano curve which

---

† There are complications (which we ignore here) when we consider the properties of the tree carefully: for example can we prevent its branches from intersecting themselves?

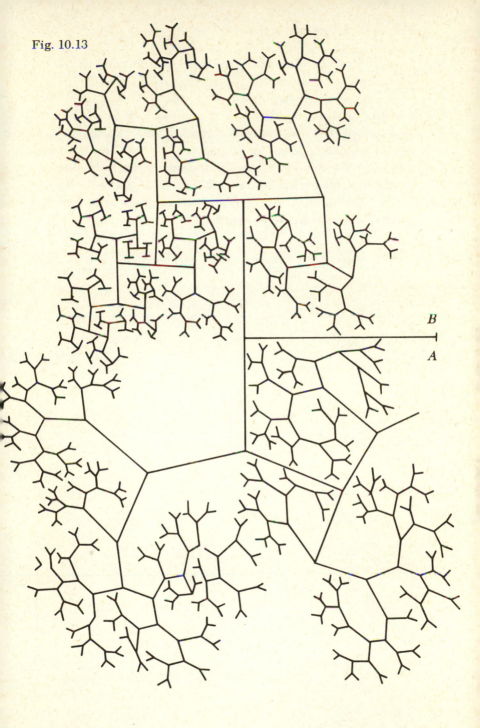

Fig. 10.13

joins *A* and *B*. Our river network thus drains its region completely.

The example of these 'unorthodox' curves shows us again that we must not be prejudiced in our attitude as to what is the 'proper business' of mathematicians. Again and again new fields have been opened up by people with new ideas and new attitudes. We should not be discouraged from following our inspiration into such fields, even if they seem strange or useless at first: the significance and 'relevance' of important new ideas is not always immediately apparent.

11

Transfinite numbers

In this book we have been discussing the concepts of 'number' and 'infinity'. It will have been apparent to the reader that, as soon as one studies these concepts in any depth, they become very closely related: compare for example our discussion of real numbers in Chapter 3. In our last chapter we discuss the notion of 'infinite' (or 'transfinite') numbers. The modern theory – one of the most beautiful and important creations in mathematics – is mainly due to Georg Cantor (1846–1918).

The creators of analysis in the nineteenth century had insisted that the lack of clarity about 'infinity' (like that about infinitesimals) was to be removed by abandoning any attempt to treat infinity as a *number*. Instead they restricted the discussion to quantities which could become 'as large as one liked', which were 'potentially', not 'actually', infinite. In a calculus course one is discouraged from writing $1/0 = \infty$, since ∞ is not a number like 5 or $\pi/2$; but one may say that $1/x \to \infty$ as $x \to 0$, which means that $1/x$ can be made larger than *any* given (finite) number N by making x small enough; note that here the 'infinitesimally small' x and the 'infinitely large' $1/x$ appear together. This caution about ∞ is essential, because the standard rules of arithmetic (see Chapter 3) apply to finite numbers only; carelessness about this point leads to the familiar mathematical paradoxes which arise when we secretly 'divide by zero'. 'Potentially infinite' collections, that is to say, finite sets which can be made larger and larger without limit, had already been considered by Aristotle who distinguished them from 'actual', 'completed', infinite sets. From ancient times, of course, examples of 'actually' infinite sets had been familiar: the collection of *all* the positive integers, say; or all the *even* positive integers; or all positive fractions; or all the points on a line, all the points in a plane, and so on. If we are to do mathematics with such infinite sets we must find a way of comparing them, so that we can decide whether the 'size' of one set is equal to, or perhaps greater than, that of another. These are old questions which have been discussed throughout history, not only by mathematicians – scholastic philosophers in the Middle Ages speculated on the nature of such infinite collections as the number of angels that could balance on the point of a needle. Galileo, in his *Dialogue on Two New Sciences* (1638), also speculates on the subject (compare Stein 1976, ch. 18). Cantor was much influenced by earlier thinking on the subject. Following work by his friend Dedekind he showed (starting around 1873) that the attitude of Cauchy and Weierstrass in rejecting 'completed' infinite sets from mathematics was over-cautious, and that it was indeed possible to give clear and rigorous rules for calculating with such sets, provided one was willing to admit that these rules may differ from those for finite sets (and may, as we shall see, put strain on one's preconceived notions as to how sets 'ought to' behave).

Cantor's basic idea which makes it possible to compare infinite sets is extremely simple, and extremely powerful. How do we compare finite sets, for example a collection of m boys and n girls? Are there more boys or more girls? With finite sets we can just *count the numbers* in each set and compare them. With infinite sets elementary counting is not possible, since the counting process never ends. But we can do something else – something even more basic – which does not require us to count. We can start a dance in which each boy partners a girl and, when they have all taken their partners, we just see whether there are any *unpaired* boys left (then the number of boys is greater than the number of girls) or any unpaired girls (then the number of girls is greater than the number of boys) or, if *everybody is paired off* and no one left unpaired, then the number of boys is exactly *equal* to the number of girls. Note carefully that, to establish this equality, we do not actually need to know *how many* boys or girls there are in the set.

But this method also works for infinite sets! Let us try it out by comparing the set of *natural numbers* 1, 2, 3, . . . , with the set of *even integers* 2, 4, 6, . . . (we consider only positive numbers but the argument easily extends to negative numbers). Can the members of these two sets be paired off, *with no member left unpaired*? The answer is, clearly, 'yes': we just have to associate 1 with 2, 2 with 4, 3 with 6, . . . , and generally n with $2n$. Thus we must conclude that the 'number' of integers 1, 2, 3, . . . , is *the same* as the number of even integers 2, 4, 6, . . . ! This number, which is clearly not finite, is our first *transfinite number*: it is denoted by $\aleph_0$ ($\aleph$ is the Hebrew letter 'aleph', so the number is called 'aleph zero'). It is not to be called 'infinity': that is too vague a concept.

The conclusion that our two sets have 'the same' number of elements is a straightforward deduction from our pairing principle, and yet it seems absurd. After all, the even integers do not include the elements 1, 3, 5, . . . (technically, the even integers form a *proper subset* of the set of all the integers), so there are surely *fewer* of them than the numbers in the set of *all* the integers! Well – it all depends on what you mean by 'fewer'! The confusions and apparent paradoxes in this subject arise from the transfer of everyday language, acquired from experience with finite collections, to infinite sets where we must train ourselves to work strictly with the mathematical rules of the game even though they lead to surprising results. As it happens often in the history of mathematics, Cantor's fellow-mathematicians when presented with his bold and unorthodox ideas were reluctant to abandon traditional ways of thought and language, and there was much opposition (this no doubt contributed to Cantor's frequent states of depression; proper recognition came to him only towards the end of his life). By now these controversies are a part of history, and Cantor's arguments have long since been accepted as not only mathematically sound but also as highly significant for the development of mathematics.

The peculiar property of infinite sets which is illustrated by our pairing of all the integers with the even integers is that *an infinite set can be 'put into one-to-one correspondence' with a proper subset of itself*. This is a general property of infinite sets which is not possessed by any finite set (convince yourself of this!), and it can be used to give a formal definition of what we mean by an infinite set. Such a definition had in fact been proposed by Dedekind in 1872 and gave impetus to Cantor's work.

Now let us look at the number of elements in some other infinite sets. How about the set of all (positive) *rational numbers* a/b (a, b positive integers)? Are there 'more of them' than $\aleph_0$? One would surely think so.... Here we have a rather more subtle problem. Unlike the sets 1, 2, 3, ..., or 2, 4, 6, ..., the rationals cannot be arranged 'in order', such that any fraction a/b is followed by the 'next bigger' one; there is no 'next bigger' one. Recall (p. 32) the property of the rationals of being *dense*: between any two rational numbers there is an infinite number of other rationals.

Nevertheless, the rationals can also be paired off with the integers, so that their 'number' is also $\aleph_0$. Any set with this property is said to be *countable* or *denumerable*. The denumerability of the rationals was demonstrated by Cantor in 1873. It is necessary to devise a way of 'counting' the fractions a/b, such that any fraction a/b is uniquely associated with one (and only one) integer. This can in fact be done in many ways; any one of them will do. We may for example arrange all the fractions in a square array as follows, with all fractions having denominator 1 in the first row, all fractions having denominator 2 in the second row, and so on:†

```
1/1  →  2/1      3/1  →  4/1      5/1  →  6/1      ...
1/2     2/2      3/2     4/2      5/2      ...
 ↓
1/3     2/3      3/3     4/3      ...
1/4     2/4      3/4     ...
 ↓
1/5     ...
```

Now we can associate each occupant of the array uniquely with an integer (1, 2, 3, ...) by following the path indicated by the arrows. (Note that it is useless to try to number off by going

† We are here treating $\frac{1}{2}$ and $\frac{2}{4}$, etc., as separate fractions. This makes our argument simpler, but it is not necessary. The proof sketched here is not Cantor's original proof, but a simpler one that he gave later.

along the whole of the first row, then the second row, and so on – why?) In this way a/b (the fraction in column 'a' and row 'b' of the table) is assigned a definite integer, depending on a and b. The reader can work out the formula for this association as an exercise, but it is not needed for the purpose of our proof: just *how* the numbering is done is not important; what is important is *that* it can be done. Thus the set of rationals is denumerable! This result is actually a special case of a general theorem: suppose we have a denumerable collection of denumerable sets $S_1, S_2, S_3, \ldots$; then the union of these sets is also denumerable. In the above example S_1 is the set of rationals with denominator 1, S_2 the set of those with denominator 2, and so on. An extension of this type of argument can be used to show (see p. 154 below) that the set of all *algebraic numbers* is also denumerable.

You might by now begin to think that all infinite sets are denumerable. If this were indeed so the subject of the size of infinite sets would not be terribly interesting, as they would all be equivalent. Towards the end of 1873 Cantor found the crucial result which established different 'orders of infinity': he proved that the set of *real numbers* is not denumerable (or, equivalently: the set of *all points on the line* is not denumerable). The proof is very simple, and very ingenious.† The idea is to show that no labelling with integers can possibly be carried out which includes all the real numbers. It is in fact sufficient to consider the real numbers in some finite interval, say $(0, 1)$, of the real axis (where the end points are excluded). If this set *is* denumerable, then we can label these real numbers with the integers, writing them as $a_1, a_2, a_3, \ldots$, and we then write each one as a *non-terminating* decimal as follows:

$$a_1 = 0.a_{11}a_{12}a_{13}\ldots,$$
$$a_2 = 0.a_{21}a_{22}a_{23}\ldots,$$
$$a_3 = 0\ a_{31}a_{32}a_{33}\ldots,$$

etc. (Note that any terminating decimal, for example 0.5, can always be written in this way by writing it as $0.4999\ldots$) Every real number has a unique decimal representation of this type

† Again, we give not Cantor's original proof, but a simpler one, discovered by Cantor in 1890.

(see p. 36). To show that any such list cannot possibly contain *all* the real numbers between 0 and 1 we now construct a real number between 0 and 1 which is not in the above list as follows: we look at the 'diagonal' digits $a_{11}, a_{22}, a_{33}, \ldots,$ (a_{nn} is the nth digit in the decimal representation of a_n). We then form digits $b_1, b_2, b_3, \ldots,$ which are such that b_1 differs from a_{11}, b_2 differs from a_{22}, b_3 differs from a_{33}, and so on; for example we can take $b_n = 9$ if $a_{nn} \neq 9$, and $b_n = 1$ if $a_{nn} = 9$. Now we use the bs so defined to construct the real number

$$b = 0.b_1 b_2 b_3 \ldots b_n \ldots .$$

Then b differs from each a_i in our (denumerated) list! It can only be equal to an a_i if *each* digit in the decimal representation of b equals the corresponding digit in the decimal representation of a_i. But b cannot equal a_1 because $b_1 \neq a_{11}$; it cannot be a_2 because $b_2 \neq a_{22}$; and so on: it cannot be a_n because the nth digit in b differs from the nth digit in a_n. Thus we have shown that our list does not contain all the real numbers.

The transfinite number to be associated with the set of real numbers between 0 and 1 is thus not $\aleph_0$; it is called c (for 'continuum'). This gives us a second transfinite number which is *greater* than $\aleph_0$. Why can we say that it is greater? An infinite set A is *smaller* than a set B if the set A is in one-to-one correspondence with a subset of B but not with B itself. The rationals between 0 and 1 form a subset of the reals, so we see that $\aleph_0 < c$. Thus Cantor's notion of pairing can be used to give a perfectly well-defined meaning to the concepts of 'equal', 'greater' and 'less' for infinite sets (and these, moreover, are consistent with the usual meaning for finite collections).

There are other ways of demonstrating that the real numbers are not denumerable. A very instructive, more 'geometrical', approach is to think of the real numbers between 0 and 1 as points on the real line and to suppose again that they are denumerable, so that we can label them $a_1, a_2, a_3, \ldots,$ and display them on the line (Fig. 11.1). Now *enclose* all these

Fig. 11.1

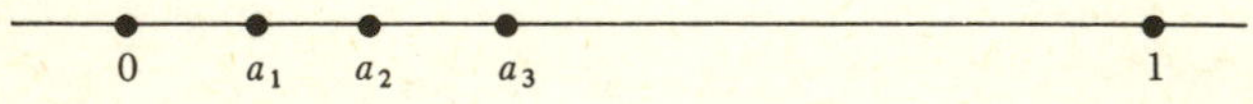

points in small intervals of the line as follows: an interval of length (say) $1/10$ is drawn containing the point a_1, an interval of length $1/10^2$ contains a_2, an interval of length $1/10^3$ contains a_3, and so on (these intervals may of course overlap, but this does not affect the argument). The *total length* of all these intervals, containing our points $a_1, a_2, a_3, \ldots$, is $1/10 + 1/10^2 + 1/10^3 + \cdots$, and the sum of this geometric series is

$$\frac{10^{-1}}{1 - 10^{-1}} = \frac{1}{9}.$$

Thus we have shown that any denumerable set of points can be enclosed in a set of intervals of total length only $\frac{1}{9}$, which indicates that such a set cannot cover the whole length of the line from 0 to 1. In fact, there is nothing special about the result $\frac{1}{9}$: we could have used, instead, intervals $\varepsilon/10$, $\varepsilon/10^2, \ldots$, with arbitrary ε, thus enclosing our denumerable set of points in a set of intervals of total length $\varepsilon/9$, and this length can be made *as small as we please* by taking ε small enough! This approach can be used to construct a well-defined way of measuring the 'length' of the interval (0, 1) occupied by a denumerable set of points (such as the rationals); such a set is said to have the *measure zero*. It shows strikingly that the set of rational numbers is a smaller infinite set than the set of reals.

The result that the set of real numbers is 'larger' than the set of rationals shows, of course, that the set of real numbers must contain numbers other than the rationals: so Cantor's approach gives us a new proof, although an indirect one, that irrational numbers exist. There are many related interesting questions which we cannot discuss here; the subject of transfinite numbers has become a major branch of mathematics. Cantor's real breakthrough was to widen the whole scope of mathematics and of the idea of 'number' in particular; he created an entire 'transfinite arithmetic' with its own rules for calculating with transfinite numbers. What other transfinite numbers are there, beyond the two ($\aleph_0$ and c) we have encountered so far? That is a deep and difficult question which has not yet been fully answered (see below), but let us note one important point. The

numbers $\aleph_0$ and c are (transfinite) *cardinal numbers*; they have
to do with a process analogous to counting the members of a set
and are similar to 'one', 'two', 'three', ... In finite arithmetic we
also have *ordinal numbers* ('first', 'second', 'third', . . .) which
tell us about the *order* of the elements in a set. In the finite case
we need not bother too much about the distinction between the
arithmetic of cardinal and the arithmetic of ordinal numbers;
there are no significant differences. In the case of infinite sets
this is no longer the case, and we have to study and define the
possibility of *ordering relations* very carefully for such sets.
This leads to a separate theory of *transfinite ordinal numbers*
which was also given by Cantor; its details are beyond our
scope. It turns out that the apparently simple notion of ordering
leads in the case of infinite sets to some of the very deepest
questions in modern mathematics. We should also stress that
the construction of number systems is by no means a finished
chapter in mathematics. A. Robinson's extension of the number
system which underlies his non-standard analysis was mentioned in Chapter 10, and a very interesting recent approach
is due to Conway (1976), whose very general construction
encompasses the real numbers, the transfinite cardinals and
ordinals and in addition many remarkable generalisations of
infinite and infinitesimal numbers.

Some questions of immediate interest in connection with the
set of real numbers are the following:

(1) What about the difference between algebraic and transcendental numbers? (Recall (p. 65) that algebraic numbers are
defined as roots of $P(x) = 0$, where $P(x)$ is a polynomial with
integer coefficients, and remember that Liouville had in 1844
constructed numbers which are not algebraic.) Now Cantor
showed that not only the rationals but the much more general set
of algebraic numbers forms a denumerable subset of the real
numbers. (To show this one proves that the set of all polynomials with integer coefficients is denumerable, and this is
done by pairing off this set with the rational numbers.) Since
the reals are not denumerable, we have here another proof
that transcendental (non-algebraic) numbers exist; and the
argument shows, furthermore, that it is the transcendental

numbers which give to the real number system the density which results in a cardinal number greater than $\aleph_0$. Transcendental numbers are not occasional curiosities: *all* numbers except those in a set of measure zero are transcendental! Unfortunately this type of existence proof does not allow us to construct any *examples* of the objects whose existence is proved; they are 'indirect', 'non-constructive' proofs. Such proofs, while perfectly logical, are in a sense not very satisfying, and attempts have been made from time to time to banish them from mathematics. But the cost is too high; it cannot be done without sacrificing large and essential aspects of mathematics.

(2) What about the cardinal number to be associated with *all* the reals, i.e. all the points on the whole infinite real line; is this greater than c? Another surprise: the cardinal of any segment of the real line is just c; it is the *density* of the points that matters, not the length of the interval. To show this we have to demonstrate a one-to-one correspondence between points of the two segments we wish to compare. This is illustrated in Fig. 11.2 for the comparison of the intervals $AB = (0, 1)$ and $AC = (0, \infty)$.

(3) An even stranger result is obtained when we ask about the number of points in a square (a *two-dimensional* set of points); surely *this* must have a cardinal number greater than c. That is certainly what Cantor believed at first, and he tried very hard to prove it, but in 1874 he found to his own amazement that he was wrong: the cardinal number is also c! (He wrote to Dedekind; 'I see it, but I don't believe it'.) The demonstration is simple enough: as always, we must establish a one-to-one correspondence, in this case between points (x, y) of the unit square (Fig. 11.3) and points on the line segment from 0 to 1. This can be done (apart from minor complications) by writing x

Fig. 11.2

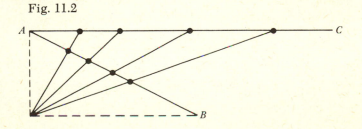

and y as decimals: $x = 0.a_1a_2a_3\ldots$, $y = 0.b_1b_2b_3\ldots$, and then associating with the point (x, y) of the square the unique point $z = 0.a_1b_1a_2b_2a_3b_3\ldots$ of the line segment (whose decimal representation is constructed by mixing up the representations of x and y in the way indicated). Clearly we can show by a similar process that the cardinal number of the points in a *cube* is also c, and so on. The cardinal numbers of sets of points in spaces of arbitrary dimension are all equal; 'cardinality' is not the same as 'dimension'. This startling result underlines again, though in a different way from that discussed in the last chapter, the difficulties involved in giving a satisfactory definition of dimension.

After all this we may now begin to wonder whether there are in fact any cardinal numbers other than $\aleph_0$ and c. The answer is *yes*! Cantor showed that: *given any set A, we can always construct a set B with a greater cardinal number.* Thus we can go on constructing ever greater cardinal numbers, and there is no greatest one. The proof is very beautiful but, as we are dealing here with a very general theorem in set theory, the definitions and arguments are necessarily very abstract. Given A, the set B is defined to be the set whose elements are *all the subsets of the given set A* (including A itself and the empty set $\varnothing$). For example, if A is the finite set $\{1,2,3\}$, then B is the set whose elements are $\{1,2,3\}$, $\{1,2\}$, $\{1,3\}$, $(2,3)$, $\{1\}$, $\{2\}$, $\{3\}$ and $\varnothing$. Note that B contains $8 = 2^3$ elements. (*Exercise*: show that, if A contains N elements, where N is any positive integer, then B

Fig. 11.3

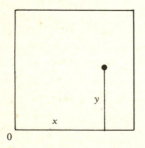

contains 2^N elements.) B is called the *power set* associated with A (we are not, of course, confining our arguments to finite sets).

We now show that there can be no one-to-one correspondence between the elements a of A and the elements of B (the subsets of A). Suppose there is such a pairing: $a \leftrightarrow S_a$, so that S_a *is the subset of A corresponding to the element a of A.* The subset S_a may or may not itself contain the element a to which S_a has been assigned. So the sets S_a are of two types: those which themselves do contain the element a (to which they are assigned), and those which do not. Now we construct a subset T of A which cannot be correlated with any element a! T *is the subset of A which consists of all those elements x of A such that S_x does not contain x.* This set differs from all the S_a above, i.e. T cannot be identified with any of the S_a. For, if S_a contains a, then T (from its definition) does not; and it cannot be an S_a which does not contain a: if S_a does not contain a, then T (from its definition) does! Hence T is not included in the above correspondence, and the proof is complete. It is easy to see that the cardinal number of B must be greater than that of A: there is a one-to-one correspondence between elements a of A and the subset of B which consists of all *one-element* subsets $\{a\}$ of A.

What happens when A and B are infinite sets? One remarkable relation is fairly easily obtained. Suppose A is the set of natural numbers, containing $\aleph_0$ elements. Then any infinite subset of A, consisting of numbers N_i such that $0 < N_1 < N_2 < N_3 < \cdots$, corresponds to a 'binary decimal'

$$\frac{1}{2^{N_1}} + \frac{1}{2^{N_2}} + \frac{1}{2^{N_3}} + \cdots,$$

and thus to some *real number* between 0 and 1. The totality of all such subsets gives the real number continuum, and thus we have the interesting equation $2^{\aleph_0} = c$ relating the transfinite cardinals $\aleph_0$ and c.

There are, however, disquieting aspects to Cantor's result that, given any set whatsoever, we can always find a bigger set. It leads us into logical difficulty. Suppose we contemplate the *set of all cardinal numbers.* Then we see that there can be no such thing, for, given this set, we can immediately construct a set with a *larger* cardinal, so that our original set cannot have contained *all* the cardinals! Many such logical paradoxes arise in set theory when we look at very general sets: they indicate that the 'naive' concept of a set, as simply any collection without restriction of any kind, is not after all so simple when it comes to infinite collections, and one must think very carefully about the 'rules of the game' in set theory if one wants to avoid contradictions. This realisation has let to an immense amount of work on the logical foundations of mathematics; it is by no means completed.

We note finally another curious point which has proved to be unexpectedly difficult. This is the question whether the cardinal number c of the real-number continuum is the 'next bigger' cardinal after $\aleph_0$. We have not assumed this (that is why we have not called this cardinal $\aleph_1$); in other words, we have left open the question whether there are infinite sets with cardinal numbers *between* $\aleph_0$ and c – or, what is the same thing, whether there are infinite sets of points on the real line which are not denumerable and which also cannot be put into one-to-one correspondence with the real numbers. This question arises naturally when one compares infinite sets, and it might seem at first sight that it should not be too difficult to settle. Cantor believed that there were no such intermediate cardinal numbers – he called this the *continuum hypothesis* – and he tried for many years to find a proof. He failed, and so has everyone else who has tried his hand. The deep-seated nature of the problem has been demonstrated by Kurt Gödel (in 1936) and Paul Cohen (in 1963). They have proved that the continuum hypothesis cannot be deduced from the standard axioms of set theory, but that, if it is adopted as a separate mathematical principle, then no contradictions will occur. These results are most remarkable. They mean that the continuum hypothesis is, in a deep sense, *unprovable*. Any decision about it involves consideration of the basic axioms to be adopted for set theory and thus for the whole of mathematics. We remain free to speculate about the 'right' form of these axioms, and whether our eventual choice will allow us to conclude that the continuum hypothesis is either true or false. These are questions about the very nature of mathematics: they remain open. The notion of infinity, so mysterious and tantalising since the days of Zeno, has not lost its power to intrigue us and to produce puzzles of deep and subtle complexity.

APPENDIX

Essay topics

As mentioned in the preface, this list is intended to help the student to pursue in greater depth something touched upon in the text, or suggested by it. A good essay should:

(i) contain some discussion of mathematical ideas (i.e. it should not be just a list of dates, or unexplained problems);

(ii) be interesting to the reader;

(iii) be comprehensible to the reader (who may be presumed to have a little mathematical ability);

(iv) give some indication that the writer understands the ideas involved;

(v) be written essentially in the writer's own words;

(vi) not complicate matters when a simple explanation is possible (thus it should not introduce unnecessary technical words).

The exact length of the essay is immaterial, but 3000–4000 words is recommended as a norm. The sources consulted should be listed at the end of the essay (the works listed in the bibliography will serve as a guide).

1 'The historical growth of a science does not necessarily pass through the stages in which we now develop it in our instruction.' Does the history of mathematics bear this out?

2 'In all the records of ancient civilisations there is evidence of some preoccupation with arithmetic over and above the needs of everyday life.'

3 'Oriental mathematics never seems to have been emancipated from the millennial influence of the problems in technology and administration, for the use of which it had been invented.'

4 'Already the pronounced tendency towards tediousness, which seems to be inherent in elementary mathematics, might plead for its late origin, since the creative mathematician would prefer to pay his attention to the interesting and beautiful problems' (Speiser). Describe some 'interesting and beautiful' mathematical problems of ancient origin, and their modern treatment.

5 'The eighteenth century had the misfortune to come after the seventeenth and before the nineteenth.' Discuss (in relation to mathematics).

6 Bertrand Russell claimed that it was the nineteenth century that discovered the nature of pure mathematics. What did he have in mind?

7 Mathematics in China and India.

8 The attitude to mathematics of Plato and Aristotle.

9 Euclid.

10 The Greeks' study of conic sections.

11 The mathematics of the Islamic period.

12 Mathematicians of the French Revolution.

13 The Bernoulli family.

14 Euler.

15 'Gauss is sometimes described as the last mathematician to know everything in his subject.'

16 Negative numbers.

17 Fractions: decimal, sexagesimal, and 'unit'.

18 The discovery of logarithms.

19 The history of the square root of minus one.

20 The solution of algebraic equations.

21 The number π.

22 Why are vectors so important in science and mathematics?

23 Squaring the circle.

24 'The method of exhaustion is a rigorous but sterile method.'

25 To what extent can the Greeks be said to have anticipated the 'infinitesimal calculus'?

26 'Seventeenth- and eighteenth-century mathematicians had little understanding of infinite series.' Discuss, with illustrations.

27 'Infinite processes were still carelessly handled in the eighteenth century and much of the work of the leading mathematicians of that period impresses us as wildly enthusiastic experimentation.' Discuss, with examples.

28 'Fermat, the true inventor of the differential calculus' (Laplace).

29 'Taking mathematics from the beginning of the world to the time of Newton, what he has done is much the better half' (Leibniz).

30 Did Newton and Leibniz 'invent' the calculus?

31 Joseph Fourier.

32 Georg Cantor.

33 What can we learn from a study of the history of mathematics?

34 'The triumph for a historian of science is to prove that nobody ever discovered anything' (Hadamard).

BIBLIOGRAPHY

The literature on the history of mathematics is immense. We give a selection of books and articles, at varying levels of difficulty, which we have found useful and which are recommended for further reading and deeper study. Books of general interest are listed first, followed by references related to particular chapters; some items additional to those mentioned in the text have been included, and some brief indications of scope and content have been given. Many further references will be found in the works listed here.

GENERAL WORKS

(i) Histories

E. T. Bell (1945) *The Development of Mathematics* (2nd edn) (McGraw-Hill, New York). (Chatty history of ideas.)

E. T. Bell (1937) *Men of Mathematics* (Gollancz, London). (Entertaining light reading, but as a history must be treated with caution.)

N. Bourbaki (1969) *Eléments d'histoire des mathématiques* (Hermann, Paris). (Collection of historical appendixes from the famous Bourbaki texts; highbrow).

C. B. Boyer (1968) *A History of Mathematics* (Wiley, New York). (Good American College text; many references.)

F. Cajori (1924) *A History of Mathematics* (Macmillan, New York). (Rather old but full of interesting items.)

F. Cajori (1928) *A History of Mathematical Notations*, Vol. 1: *Notations in Elementary Mathematics* (The Open Court Publishing Co., La Salle, Ill.) (Reprinted 1974).

P. Dedron & J. Itard (1978) *Mathematics and Mathematicians 1 and 2* (Open University Press, Milton Keynes). (Set book for Open University course.)

J. Dieudonné *et al.* (1978) *Abrégé d'histoire des mathématiques 1700–1900 I and II* (Hermann, Paris). (Authoritative 'digest' at advanced undergraduate standard.)

J. M. Dubbey (1970) *Development of Modern Mathematics* (Butterworths, London).

H. Eves (1969) *An Introduction to the History of Mathematics* (3rd edn) (Holt, Rinehart and Winston, New York). (Good for earlier history and elementary mathematics.)

J. E. Hofmann (1963) *Geschichte der Mathematik I, II and III* (Sammlung Göschen: De Gruyter, Berlin). (Extensive list of sources; indispensable for the serious historian.) (English translation, without sources, published as *The History of Mathematics to 1800* by Littlefield, Adams and Co., Totowa, N.J., 1967).

M. Kline (1972) *Mathematical Thought from Ancient to Modern Times* (Oxford University Press, New York). (Useful for reference.)

D. J. Struik (1954) *A Concise History of Mathematics* (Bell, London). (Masterly summary, quite highbrow. Emphasises social aspects.)

(ii) Ideas and concepts

E. T. Bell (1952) *Mathematics: Queen and Servant of Science* (Bell, London).

R. Courant & H. Robbins (1941) *What is Mathematics?* (Oxford University Press, London). (A superb survey.)

T. Dantzig (1947) *Number, the Language of Science* (Allen and Unwin, London). (A stimulating essay.)

H. Eves & C. V. Newsom (1965) *An Introduction to the Foundations and Fundamental Concepts of Mathematics* (Holt, Rinehart and Winston, New York).

H. B. Griffiths & P. J. Hilton (1970) *A Comprehensive Textbook of Classical Mathematics: A Contemporary Interpretation* (Van Nostrand, London). (Technical survey of 'modern mathematics', based on a course for school-teachers.)

T. S. Kuhn (1970) *The Structure of Scientific Revolutions* (University of Chicago Press). (A seminal work in the history of scientific ideas.)

I. Lakatos (1976) *Proofs and Refutations* (Cambridge University Press). (Profound and entertaining discussion of the development of mathematical ideas.)

G. Polya (1948) *How to Solve It* (Princeton University Press). (Shows how the mathematician sets about solving a problem.)

H. Rademacher & O. Toeplitz (1957) *The Enjoyment of Mathematics* (Princeton University Press). (Analysis of interesting problems for the non-specialist.)

W. W. Sawyer (1955) *Prelude to Mathematics* (Penguin Books, Harmondsworth), and

W. W. Sawyer (1970) *The Search for Pattern* (Penguin Books, Harmondsworth). (Good popular works.)

S. K. Stein (1976) *Mathematics, the Man-Made Universe* (3rd edn) (Freeman, San Francisco). (A selection of topics at a level similar to ours.)

S. J. Taylor (1970) *Exploring Mathematical Thought* (Ginn, London). (Concise introduction to important ideas and to mathematical notation.)

(iii) Source books and anthologies

H. Midonick (1965) *The Treasury of Mathematics* (Peter Owen, London).

J. R. Newman (1956) *The World of Mathematics*, vols. 1–4 (Simon and Schuster, New York).

D. J. Struik (1969) *A Source Book in Mathematics, 1200–1800* (Harvard University Press, Cambridge, Mass.).

(iv) Journals
Historia Mathematica (Academic Press, New York), and
Archive for History of Exact Sciences (Springer, New York). (Valuable
  sources of articles on special topics.)

CHAPTER 1

J. M. Dubbey (1978) *The Mathematical Work of Charles Babbage*
  (Cambridge University Press).
C. H. Edwards (1979) *The Historical Development of the Calculus*
  (Springer, New York). (See ch. 6 for a fuller account of the introduction of
  logarithms than is given here.)
H. H. Goldstine (1972) *The Computer from Pascal to von Neumann*
  (Princeton University Press).
T. L. Heath (1921) *A History of Greek Mathematics*, vols 1 and 2
  (Clarendon Press, Oxford). (The standard work.)
O. Neugebauer (1969) *The Exact Sciences in Antiquity* (Dover, New York).
  (An account of ancient mathematics, by one of the leading interpreters.)
H. Rademacher & O. Toeplitz (1957) *The Enjoyment of Mathematics*
  (Princeton University Press).
B. L. van der Waerden (1954) *Science Awakening* (Noordhoff, Groningen).
  (Very readable account of ancient and Greek mathematics.)
C. Zaslavsky (1973) *Africa Counts* (Prindle, Weber and Schmidt, Boston,
  Mass.) (African number systems.)

CHAPTER 2

H. Davenport (1970) *The Higher Arithmetic* (Hutchinson, London).
  (Readable introduction to the theory of numbers; many useful
  references.)
U. Dudley (1978) *Elementary Number Theory* (2nd edn) (Freeman, San
  Francisco).
H. M. Edwards (1977) *Fermat's Last Theorem* (Springer, New York). (See
  especially the first three chapters; contains much interesting history.)
T. Hall (transl. A. Froderberg) (1970) *Carl Friedrich Gauss* (MIT Press,
  Cambridge, Mass.). (Short account of life and mathematics.)
M. E. Hellman (1979) The mathematics of public-key cryptography.
  Scientific American **241** (no. 2), 130–9 (August 1979).
G. B. Kolata (1980) Testing for primes gets easier. *Science* **209**, 1503–4.
D. J. Newman (1980) Simple analytic proof of the prime number theorem.
  The American Mathematical Monthly **87**, 693–6.
S. R. Ranganathan (1967) *Ramanujan, the Man and the Mathematician* (Asia
  Publishing House, London).
E. J. Scourfield (1979) Perfect numbers and Mersenne primes. *Mathematical
  Spectrum* **12**, 84–92.

CHAPTER 5

F. Cajori (1924) *A History of Mathematics* (Macmillan, New York).

CHAPTER 6

A. Baker (1975) *Transcendental Number Theory* (Cambridge University Press, London). (An advanced text.)

P. Beckmann (1971) *A History of π* (St Martin's Press, New York).

R. Courant & H. Robbins (1941) *What is Mathematics?* (Oxford University Press, London)

H. Davenport (1970) *The Higher Arithmetic* (Hutchinson, London).

O. Ore (1957) *Niels Henrik Abel* (University of Minnesota Press, Minneapolis).

I. Stewart (1973) *Galois Theory* (Chapman and Hall, London).

CHAPTER 7

M. J. Crowe (1967) *A History of Vector Analysis* (University of Notre Dame Press). (Non-technical history of vector analysis and its principal protagonists.)

R. W. Feldmann (1962) History of elementary matrix theory. *The Mathematics Teacher* **55**, 482–4, 589–90, 657–9.

W. R. Hamilton (1967) *The Mathematical Papers of Sir William Rowan Hamilton*, vol. 3: *Algebra* (ed. H. Halberstam & R. E. Ingram) (Cambridge University Press).

H. Kennedy (1979) James Mills Peirce and the cult of quaternions. *Historia Mathematica* **6**, 423–9. (Gives references to papers on the quaternion/vector controversy.)

P. M. Morse & H. Feshbach (1953) *Methods of Theoretical Physics*, part I, ch. 1 (McGraw-Hill, New York). (Spinors, quaternions and special relativity.)

L. Nový (1973) *Origins of Modern Algebra* (Academia, Prague). (Surveys developments in the first half of the nineteenth century.)

D. Quadling (1979) Q for quaternions. *Mathematical Gazette* **63**, 98–110. (Quaternions and rotations.)

CHAPTER 8

C. B. Boyer (1968) *A History of Mathematics* (Wiley, New York).

E. J. Dijksterhuis (1957) *Archimedes* (Humanities Press, New York).

O. Toeplitz (1963) *The Calculus – A Genetic Approach* (University of Chicago Press). (Historical introduction to the ideas of the calculus.)

CHAPTER 9

C. B. Boyer (1959) *History of the Calculus* (Dover, New York).

J. M. Dubbey (1978) *Development of Modern Mathematics* (Butterworths, London). (See chs. 2 and 3 for British mathematics 1800–30 and the Analytical Society.)

C. H. Edwards (1979) *The Historical Development of the Calculus* (Springer, New York). (Gives a fuller account of many topics treated here.)

G. H. Hardy (1949) *Divergent Series* (Clarendon Press, Oxford). (See pp. 18–20 for remarks on British analysis in the early nineteenth century.)

M. S. Mahoney (1973) *The Mathematical Career of Pierre de Fermat* (Princeton University Press).

CHAPTER 10

I. Grattan-Guinness (1972) *Joseph Fourier 1768–1830* (MIT Press, Cambridge, Mass.).

H. J. Keisler (1976) *Elementary Calculus* (Prindle, Weber and Schmidt, Boston, Mass.). (A 'non-standard' first course.)

B. B. Mandelbrot (1977) *Fractals* (Freeman, San Francisco). (Many examples of non-differentiable curves and other irregular configurations, with excellent illustrations.)

A. Robinson (1966) *Non-standard Analysis* (North-Holland, Amsterdam). (Highly technical, with a historical chapter.)

N. Ya. Vilenkin (1968) *Stories about Sets* (Academic Press, New York). (Entertaining account of the 'surprises' in set theory.)

CHAPTER 11

P. J. Cohen (1966) *Set Theory and the Continuum Hypothesis* (W. A. Benjamin, New York). (A technical but very clear account.)

J. H. Conway (1976) *On Numbers and Games* (Academic Press, London). (Highly technical.)

J. W. Dauben (1979) *Georg Cantor* (Harvard University Press, Cambridge, Mass.). (Full account of life, work and ideas.)

D. E. Knuth (1974) *Surreal Numbers* (Addison-Wesley, Reading, Mass.) (Entertaining popular introduction to Conway's system.)

S. K. Stein (1976) *Mathematics, the Man-Made Universe* (3rd edn) (Freeman, San Francisco).

INDEX

abacus, 6, 13
Abel, N. H., 63
  Abel–Ruffini theorem, 63
Accademia dei Lincei, 118
Achilles paradox, *see* Zeno of Elea
Adams, J. F., 78
addition, 27
  commutative law, 27
aleph zero, 149
Alexandria, 5
algebra
  Boolean, 75
  Clifford, 79
  division, 74, 78
  linear associative, 79
  non-commutative, 74
  vector, 82
Al-Khowarizmi, 6
  Al-jabr wa'l-muqabalah, 6
Analytical Society, 125
angle, trisection of, 64, 89
Antiphon, 92
Apollonius of Perga, 5, 112
Arab mathematics, 6
Archimedes of Syracuse, 5, 45, 95f
  axiom of, *see* continuity, axiom of
  estimate of π, 97
  Method, 102
  quadrature of parabola, *see* area
    of parabola
  Sand-Reckoner, 96
  spiral of, 96, 112
area
  of circle, 86, 92
  under hyperbola, 109
  of parabola, 98f
Argand, J. R., 48
  Argand diagram, *see* complex
    plane
Aristotle, 33, 147
astronomy, 4
atomism, 87
axioms, 30

Babbage, Charles, 14
  analytical engine, 14
Babylonian mathematics, 3, 42
Babylonian number system, 3f, 86
Barrow, Isaac, 117, 119, 121
  Geometrical Lectures, 119
base 2, 7, 10
base 3, 10
base 4, 7
base 10, 1
Berkeley, George, 124
  Analyst, 124
Bernoulli, Daniel, 132
Bernoulli, Jakob, 124
Bernoulli, Johann, 46, 124
Bhaskara, 45
binary number system, *see* base 2
biquadratic equation, *see* quartic
  equation
Bolzano, Bernhard, 140
Bombelli, R., 55
  Algebra, 55
Boole, George, 75
  *Investigation of the Laws of
    Thought*, 75
  *The Mathematical Analysis of
    Logic*, 75
Brahe, Tycho, 105
branching networks, 144
Briggs, Henry, 13
Brownian motion, 142
Bürgi, Jobst, 12

calculating machine, 13
calculus
  differential, 87, 111f, 123
  fundamental theorem of, 115
  integral, 87, 105f, 123
Cantor, Georg, 147f
Cardan, Jerome, 46, 52
  Ars Magna, 52
  Cardan's solution of the cubic
    equation, 51